ACE YOUR MIDTERMS & FINALS

INTRODUCTION TO PSYCHOLOGY

Other books in the Ace Your Midterms and Finals Series include:

Ace Your Midterms and Finals: Introduction to Biology

Ace Your Midterms and Finals: U.S. History

Ace Your Midterms and Finals: Fundamentals of Mathematics

Ace Your Midterms and Finals: Introduction to Physics

Ace Your Midterms and Finals: Principles of Economics

ACE YOUR MIDTERMS & FINALS

INTRODUCTION TO PSYCHOLOGY

ALAN AXELROD, PH.D.

McGraw-Hill
New York San Francisco Washington, D.C. Auckland Bogotá
Caracas Lisbon London Madrid Mexico City Milan
Montreal New Delhi San Juan Singapore
Sydney Tokyo Toronto

Library of Congress Catalog Card Number: 99-070504

McGraw-Hill

A Division of The ***McGraw-Hill*** *Companies*

1 2 3 4 5 6 7 8 9 0 DOC/DOC 9 0 9 8 7 6 5 4 3 2 1 0 9

ISBN 0-07-007007-5

The sponsoring editor for this book was Barbara Gilson, the editing supervisor was Maureen B. Walker, the designer was Stateless Design for The Ian Samuel Group, Inc., and the production supervisor was Tina Cameron.

Printed and bound by R. R. Donnelley & Sons Company.

This book is printed on recycled, acid-free paper containing a minimum of 50% recycled, de-inked fiber.

HOW TO USE THIS BOOK

YOU KNOW THE DRILL. ON THE FIRST DAY IN A SURVEY, INTRODUCTORY OR "CORE" course, the professor talks about grading and, saying something about the value of the course in a program of "liberal education," declares that what he or she wants from her students is original thought and creativity and, above all, he or she does not "teach for" the midterm and final.

Nevertheless, the course certainly *includes* one or two midterms and a final, and these account for a very large part of the course grade. Maybe the professor can with a straight face disclaim *teaching for* these exams, but few students would deny *learning for* them.

True, you know that the purpose of an introductory course is to gain a useful familiarity with a certain field, not just to prepare for and do well on a two or three exams. Yet the exams *are* a big part of the course, and, whatever you learn or fail to learn in the course, your performance as a whole is judged in large measure by your performance on these exams.

So the cold truth is this: More than anything else, curriculum core courses *are* focused on the midterm and final exams.

Now, traditional study guides are outlines that attempt a bird's-eye view of a given course. But *Ace Your Midterms and Finals: Principles of Psychology* breaks with this tradition by viewing course content through the magnifying lens of ultimate accountability: the course exams. The heart and soul of this book consists of eleven midterms and chapters containing finals prepared by *real* instructors, teaching assistants, and professors for *real* students in *real* schools.

Where did we get these exams? Straight from the professors and instructors themselves.

- All exams are real and have been used in real courses.
- All exams include critical "how-to" tips and advice from the creators and graders of the exams.
- All exams include actual answers.

Let's talk about those answers for a minute. In most cases, the answers are actual student responses to the exam. In some cases, however, the instructors and professors have created "model" or "ideal" answers. Usually, the answers included are A-level responses. Sometimes, however, they are not perfect (because they are real). In all cases, you'll find full commentary by the instructors, who point out what works (and why) and what could use improvement (and why—as well as how to improve it).

This book also contains more than the exams themselves.

- In Part One, "Preparing Yourself," you'll find how-to guidance on what Psychology professors look for, how to think like an psychologist, how to study more effectively, and how to gain the performance edge when you take an exam.
- Part Two, "Study Guide," presents a quick-and-easy overview of the content of typical surveys of psychology. It clues you in on what to expect in these courses.
- Part Three, "Midterms and Finals," give the exams themselves, grouped by college or university.
- In Part Four, "For Your Reference," you'll find a handy glossary of key terms in Psychology and a brief list of recommended reading.

What This Book *Is Not*

Ace Your Midterms and Finals: Principles of Psychology offers a lot of help to see you through to success in this important course. But (as you'll discover when you read Part One) the book *cannot* take the place of

- Doing the assigned reading
- Keeping up with your work and study
- Attending class
- Taking good lecture notes

- Thinking about and discussing the topics and issues raised in class and in your books

Ace Your Midterms and Finals: Principles of Psychology is not a substitute for the course itself!

What This Book *Is*

Look, it's both cynical and silly to invest your time, brainpower, and money in a college course just so that you can ace a couple of exams. If you get A's on the midterm and final, but come away from the course having learned nothing, you've failed.

We don't want you to be cynical or silly. The purpose of introductory, survey, or "core" courses is to give you a panoramic view of the knowledge landscape of a particular field. The primary goal of the college experience is to acquire more than tunnel intelligence. It is to enable you to approach whatever field or profession or work you decide to specialize in from the richest, broadest perspective possible. College is education, not just vocational training.

We don't want you to "study for the exam." The idea is to study for "the rest of your life." You are buying knowledge with your time, your brains, and your money. It's an expensive and valuable commodity. Don't leave it behind you in the classroom at the end of the semester. Take it with you.

But even the most starry-eyed idealist can't deny that midterms and finals are a big part of intro courses and that even if your ambitions lie well beyond these exams (which they should!), performing well on them is necessary to realize those loftier ambitions.

Don't, however, think of midterms and finals as hurdles—obstacles—you must clear in order to realize your ambitions and attain your goals. The exams are there. They're real. They're facts of college life. You might as well make the most of them.

Use the exams to help you focus your study more effectively. Most people make the mistake of confusing *goals* with *objectives*. Goals are the big targets, the ultimate prizes in life. Objectives are the smaller, intermediate steps that have to be taken to reach those goals.

Success on midterms and finals is an objective. It is an important, sometimes intimidating, but really quite doable step toward achieving your goals. Studying for—working toward—the midterm or final is *not* a bad thing, as long as you keep in mind the difference between objectives and goals. In fact, fixing your eye on the upcoming exam will help you to study more effectively. It gives you a more urgent purpose. It also gives you something specific to set your sights on.

And this book will help you study for your exams more effectively. By letting you see how knowledge may be applied—immediately and directly——to exams, it will help you acquire that knowledge more quickly, thoroughly, and certainly. Studying these exams will help you to focus your study in order to achieve success on the exams—that is, to help you attain the objectives that build toward your goals.

—*Alan Axelrod*

CONTRIBUTORS

Alan J. Beauchamp, *Associate Professor of Psychology, Northern Michigan University*

Terry D. Blumenthal, *Associate Professor of Psychology, Wake Forest University*

Jeffrey A. Gibbons, *Visiting Assistant Professor of Psychology, Carthage College*

Bryon Gibson, *Associate Professor of Psychology, Central Michigan University*

Fred Heilizer, *Associate Professor of Psychology, DePaul University*

Mark A. Lumley, *Associate Professor, Wayne State University*

Hajime Otani, *Professor of Psychology, Central Michigan University*

Debra Ann Poole, *Professor of Psychology, Central Michigan University*

Jeffrey J. Sable, *Graduate Teaching Assistant in Psychology, Kansas State University*

Nayantara Santhi, *Graduate Teaching Assistant in Psychology, Northeastern University*

Guy Vitaglione, *Instructor in Psychology, Kansas State University*

ABOUT THE AUTHOR

Alan Axelrod, Ph.D. , is the author of numerous books, including *Booklist* Editor's Choice *Art of the Golden West, The Penguin Dictionary of American Folklore,* and *The Macmillan Dictionary of Military Biography*. He lives in Atlanta, Georgia.

PART ONE

PREPARING YOURSELF

CHAPTER 1

PRINCIPLES OF PSYCHOLOGY: WHAT THE PROFESSORS LOOK FOR

PSYCHOLOGY IS THE SCIENTIFIC STUDY OF BEHAVIOR AND MENTAL PROCESSES. PUT less formally, psychology takes as its field the subject of what makes us tick: what motivates and drives us, how we think, how we feel, what happens when we're awake, what happens when we're asleep, how we behave alone and in groups, how we develop, how we persuade and are persuaded, how we learn, and what happens when mental processes go haywire.... The list is almost endless.

The Introductory Psychology course is intended to present (a) basic psychological methodology and terminology, (b) a broad variety of psychological knowledge and theory, and (c) integration of various approaches and areas of psychology. As a result of having taken—and passed—this course, you can expect to have a pretty good knowledge of basic psychology and its many, many, many ramifications.

—Fred Heilizer, Associate Professor, DePaul University

Psychology investigates the subject with which we are on most intimate terms: ourselves. And it investigates the subject most mysterious and bewildering to us: ourselves.

No wonder so many students eagerly sign up for introductory psych courses! After all, nothing interests us more than ourselves, our thoughts, our feelings, our motivations. And the truth is that few introductory college courses have more potential than Psych 101 to be fascinating and rewarding.

Yet too many intro students soon find themselves disappointed.

Take another look at the definition that is the first sentence of this chapter: "Psychology is the scientific study of behavior and mental processes." The key adjective is *scientific.*

By any measure, psychology is a vast, ambitious, and varied field. It confronts a subject of inquiry that is often difficult to define, to contain, and to quantify. How do you *measure* a feeling, a thought? And yet that is precisely the approach psychology takes. As a *scientific* study, psychology embraces the *scientific* method, which is explained in Chapter 4. The scientific method attempts to define, contain, and measure, and, using this method, psychologists look for ways to approach their subject such that they can define, contain, and measure feelings, behavior, actions, and other phenomena of mind. Students who enter an introductory psychology course expecting sweeping philosophical statements or hoping to be handed the keys that unlock the secrets of all human behavior will be disappointed.

I want to create in my students awareness of the breadth of psychological fields and of psychology as a science . . . an awareness of the connection between personal experiences and psychological science.

—Guy Vitaglione,
Instructor, Kansas State University

Psychology typically proceeds by small, patient steps: a problem or area of inquiry is defined, a hypothesis is made, an experiment is devised to test the hypothesis. Then what is the result? Taken together, the result of a number of experiments, studies, or observations may be a theory concerning some aspect of behavior or mental process—a theory that takes its place beside other theories. If you come into a psychology class looking for the Right Answer, you will also be disappointed. Psychology is a growing—perhaps evolving—collection of theories: scientific points of view on the mind, on behavior, and on the relation of mind to body.

The main goal of the course is to present a broad overview of psychology, introducing students to psychology as a scientific discipline. I would like students to understand how psychologists use scientific methods to investigate various phenomena. I also want them to learn various terms used in psychology.

—Hajime Otani,
Professor, Central Michigan University

There was a time when psychologists held onto and passionately defended their own particular favored theory. Psychoanalysts argued with behaviorists, for example, and humanists argued with both. Today, however, most psychologists take a more eclectic approach, freely moving about and among the various theories and schools of psychological thought, choosing from each whatever ideas seem to work best in a particular situation. This means that you will have to get used to and be comfortable with the idea of learning about a number of competing and even contradictory theories—and no one will be there to tell you which one is "right."

My primary objectives in teaching this course are

1. **To get my students to think and then think critically.**
2. **To get my students interested in psychology.**
3. **To challenge my students and make them better students.**

—Jeffrey A. Gibbons,
Visiting Assistant Professor,
Carthage College

This does not mean that psychology lacks its share of hard, right-or-wrong information. You will find no shortage of new terminology and concepts to take in, and while psychology offers few *definitive* answers, you will be asked, on midterms and finals, many *definitive* questions. Most exams in introductory psychology courses are largely or even exclusively multiple choice, true-false, and short answer (fill-in-the-blank type). If the intro courses are typically short on absolute, final answers, they are plenty long on information to absorb and commit to memory.

To avoid disappointment, then, and to perform well in the classroom and on exams, you will need to:

- Keep an open but critical mind.

- Resist the temptation to settle on any one theory as definitive.
- Live with multiple points of view on any particular issue.
- Engage material actively. (Don't passively jot down the particulars of this or that theory; think about it; question it; weigh it against other interpretations.)
- Learn new terms and concepts—develop your memory.

My primary objectives in teaching this course are to introduce students to the core areas within the field, to show them how we do research, and to get them to think about how the various topics discussed interrelate. I want my students to master basic terminology and concepts, to be able to extend and apply this learning to novel situations, and to show a good understanding of how the different focuses within psychology can be related to produce a better understanding of human behavior and cognition.

To answer the questions in my exams requires mastery of materials and terminology, an understanding of interrelationships between domains of study, and an ability to apply this understanding within the context of the question. Prepare yourself by going to class, reading the textbook, asking yourself questions as you read, and asking in-class questions of the instructor. Try to challenge the instructors' thinking whenever possible. Also, think about how the different areas of psychology relate to one another.

Master materials not only by reading and encoding key terms and passages, but by thinking about how the material relates to your own life. For example: Am I classically conditioned to behave in a particular way in a given situation? Do I selectively forget bad things and remember good things? Also, think about how the domains studied interrelate; for example, how does the study of selective attention relate to the theory of defense mechanisms put forth by Freud?

As for these exams, before you answer the question write an outline to help organize your thoughts.

—Alan J. Beauchamp,
Associate Professor,
Northern Michigan University

Is "Introductory Psychology" Even Possible?

Look at the table of contents and, in particular, the titles of the chapters in Part Two. You will notice that psychology is hardly a *single* discipline. It includes areas that seem more like biology or even medicine than psychology. It includes work with statistics that seems more appropriate to a course in math or probability than psychology. It includes experimental work that you might expect to find in a zoology lab rather than a psych course. It includes material about learning, which might be taught in an education class. It includes work with attitude and persuasion that you might find in a course on marketing or even political science. It includes information on such issues as drug abuse and sociopathic behavior that would be appropriate in a criminal justice seminar. It explores issues of prejudice and discrimination that you could find in a sociology class. It makes excursions into the realms of classic literature, mythology, and religion. It explores "strange stuff" like hypnosis and relaxation techniques.

In short, psychology is everywhere.

And that is precisely the point. One of the chief things your instructor wants to *introduce* you to is the *holistic* nature of psychology. Most psychologists do indeed specialize in a particular branch of psychology, but they do so from a perspective of knowledge of the entire field. Human beings are complex. Psychologists believe that even if you choose to study a narrowly focused aspect of human behavior, you cannot fully appreciate it without a broad understanding of its context.

In this sense, an introductory psychology course is not just a *beginning* course in psychology. It gets to the very heart of the discipline. Your instructor expects you actively to engage the multiple perspectives psychology offers.

Now, such engagement can be exciting, but it can also be overwhelming. Because of the large area covered, introductory psych courses usually move fast, and they often seem to jump from one area to another. In the introductory-level courses of many disciplines, course material is presented in building-block fashion. You learn a particular fact or principle in one lecture, and the next lecture builds on that, and so on through the semester. Introductory-level psychology, however, is more accurately described as a survey rather than as a process of building. It doesn't so much ask you to stack one concept upon another as it asks you to take in a vast array of apparently independent, but ultimately interrelated, concepts. Especially because the survey moves quickly, it is very important that you attend lectures, ask questions, and keep up with the reading. If you blink, you could miss an entire system of mind or a psychological specialty!

The primary goal of this course is to familiarize students with the wide variety of subdisciplines within the field, and to ensure that they know the *basic* findings in those subdisciplines. I want my students to be able to converse knowledgeably on the main findings within the subdisciplines. For example, after the course, they should be able to identify Piaget's stages and the main accomplishments in each stage. Another example: They should be able to identify the different theoretical approaches to the study of personality. They need to have a foundation of basic psychological knowledge.

My philosophy of introductory psychology is that it should provide the student with a foundation of psychological knowledge that can be applied in future classes. Some professors focus on issues such as critical thinking; however, I believe that before one can think critically on a given topic, one must first have a background of knowledge in the topic. Therefore, I use the introductory psychology course to help provide that foundation.

— Bryon Gibson,
Associate Professor,
Central Michigan University

Great Expectations

The focus and approach of the introductory psychology courses offered in different colleges and universities—and even by different instructors within the same department—do vary, but you can count on most introductory courses covering the material you will find in Part Two. Subject to more variety, however, is what individual instructors expect from their students.

Few instructors expect that students will enter the course with any background or preparation in psychology. After all, psychology is a subject taught in few high schools. This does not necessarily mean, however, that instructors assume everyone starts at the same place. Some instructors expect that you will come to the course with a certain degree of preparation in the sciences. This can pose a problem if you are, in fact, wholly unfamiliar with the scientific method and its assumptions. Take time, early on, to understand the meaning of the scientific method and how psychologists create theory from empirical observation.

Many instructors approach the introductory psychology survey as a body of information to be absorbed. This is evident from the popularity of multiple-choice exam formats throughout these courses, and it is especially the case in lecture-oriented courses. Other instructors, however, put more emphasis on discussion, outside reading, and even some independent research. It will help you get in synch with the course if, early on, you understand the instructor's approach to the material. Realize that, while some instructors want you to memorize and master extensive amounts of prescribed materials, others are more interested that you are learning basic principles, which you are then expected to apply to practical problems. Yet even the instructors

The class coverage represents my effort to organize and extend the text's coverage. Therefore, good class notes represent the best study outline available to the student. In addition, I strongly urge students to develop their own version of speed-writing for use in this class and others, and I illustrate speed-writing technique throughout the quarter.

—Fred Heilizer,
Associate Professor, DePaul University

who emphasize memorization of factual material appreciate (and reward) initiative. This may be particularly true among psychologists, who tend to value a combination of open mindedness and skepticism that is driven by curiosity. In any science class, whether it is a "hard" science like physics or a "softer" discipline such as psychology, it helps to be a self-starter rather than someone who expects to be spoon-fed the course contents.

It is helpful to join a study group. Explain the material to anyone who will listen, and encourage the listener to ask questions and to try to stump you. Read the assigned material before class, then read it again after class.

—Terry D. Blumenthal,
Associate Professor,
Wake Forest University

In addition to attending class, taking lecture notes, reading the textbook, and doing all assigned exercises, your instructor expects you to

- Think concepts through.
- Pay close attention to discussion of key experiments; think about what experimental results imply about theoretical assumptions.
- Understand the role, value, and limitations of experiment and other empirical methods.
- Think in terms of causes and effects—but don't jump to conclusions about cause-and-effect relationships; understand the difference between correlation and cause and effect (see Chapter 8).
- Ask questions.
- Challenge your instructor.

Use flash cards to study the material focused on in lecture. Don't assume that because you understand the material when you read it or hear a lecture on it you will remember the terms associated with the material. On the first day of class I give an hour-and-a-half lecture on how to study, which includes material on the use of the quiz hints, flash cards, studying with other class members, quizzing oneself, study time, time management, and so on.

I emphasize the difference between college and high school, and how much more study time is necessary to succeed in college. I also have a specific weekly timetable I suggest students follow: quizzes are Thursday afternoon; after taking the quiz, I suggest they go to the course Web site and get the hints for the following week's quiz; make flash cards on Thursday night and Friday (this gives a chapter 'preview'). Read the chapters on Saturday and Sunday. Study flash cards on Monday and Tuesday. Go over class notes on Wednesday. Thursday morning, you should study the flash cards until you can get them 100 percent correct.

—Bryon Gibson,
Associate Professor,
Central Michigan University

Stand on Your Head

Your bedroom. Your dorm room. Few places are more familiar to you. Did it ever occur to you to look at your room from a new perspective? Try standing on your head.

One of the most invigorating and exciting aspects of psychology is that it asks us to look at very familiar things from new perspectives. Every minute of every day, we make certain assumptions about how people perceive us and others, how people behave, how people feel, how people think. Psychologists stop to question these assumptions. They question those things most of us take for granted. This is not done for the sake of being ornery or investigating the obvious, but for learning about the truths that escape us precisely because they are so often right in front of our faces.

If you try to get into the habit of considering the ordinary from extraordinary perspectives, you will soon find yourself very much in tune with the objectives of psychology in general and of the introductory course in particular.

Accordingly, you should work to develop your powers of observation and your skill at attending to detail. Moreover, you need a willingness to entertain new ideas and, while absorbing the details, you need to look beyond them to the "big picture." Psychology is a science of *systems*—parts functioning within wholes. Get into the habit of learning how units function and work together in complete systems.

Memorization and Study

Even for students who are fascinated from the start by the scope and goals of psychology, the introductory course can present some formidable challenges. Every aspect of psychology involves learning new terms—many of them of a technical nature, some of Latin or Greek derivation (difficult to pronounce, spell, and memorize), and many of them similar enough to cause confusion.

- Commit yourself to memorizing plenty of these terms.
- Associate a term with a specific example. Don't try to learn a list of terms in a vacuum.
- Similarly, it is best to learn associated terms together. For example, the time to learn the meanings of *dendrite, axon, myelin sheath,* and *synapse* is when you are studying nerve cells and are learning how these structures work together as a system.
- Identify and pay special attention to the *key* terms. Your instructor will point out many of these. Obviously, too, if a term is used often in lecture or in your textbook, you must regard it as a key term, and you should become comfortable with it.
- The best way to learn strange new terms is to make them familiar by actually using them. Practice.

The study habits that work for other college courses work well for psychology, too. Because each lecture introduces new concepts—and a set of new terms—it is very important to:

- Read textbook assignments *before* lectures.
- Keep up with your reading.
- Attend the lectures.
- Take good notes.

Course objectives include:

- **Helping students to become familiar with basic history, theories, and procedures of psychology**
- **Giving students the ability to generalize and apply psychological concepts to the 'real world'**
- **Encouraging students to think critically about the practice and implications of psychology**
- **Helping students to articulate their ideas about psychology**

I hope that the students will be able to see how psychology is relevant in some way to just about every aspect of their lives, and that they will be able to apply the concepts they learn in class to their own lives. I also believe it is very important for students to think and respond critically to what they see and hear about psychology. I want them to question the things they see and hear and not just blindly accept them.

The course contains a great deal of varied material.
If you have any question whatsoever, ask it. Many introductory psychology courses have students from very diverse academic perspectives. It is very difficult to present effectively such a diverse subject as psychology to such a diverse group of students without some feedback. I have always found that productive questions can really facilitate understanding. Often, if one person has a question, several other students have the same question.

Try to see how each subject relates to something that interests you. Not only will you learn and remember more, but the information will also be more useful to you.

— Jeffrey J. Sable,
Graduate Teaching Assistant,
Kansas State University

I would like the students to be able to understand and think through the various concepts they learn in the course. Apart from factual questions, I want them to be able to answer conceptual and integrative questions regarding the material taught in the course. The course examinations consist of true-or-false and multiple-choice questions. While some of the questions in the exam may involve facts, most of them are conceptual in nature and require a good understanding of the material. Accordingly, students should keep up with the material in class. Usually an introductory course involves a lot of material; therefore, students should learn the material as soon as it is taught in class. They should try to understand the concepts rather than merely memorize facts. I would also advise them to use study guides and do the sample exams in the guides. Students should develop an organized schedule for efficient learning. It should be consistent and systematic rather than cramming before the exam. It is important that students read the text in detail. They should take effective notes during the lecture and use them as a guide to selecting material to focus on.

—Nayantara Santhi,
Graduate Teaching Assistant,
Northeastern University

Visual Learning

Psychology includes a good many graphs, diagrams, and other graphic representations. When you encounter material like this in your textbook and in lecture presentations, remember that these are more than decorative illustrations. They are important tools for understanding major concepts. One good way to study key diagrams is to try drawing from memory at least the major features. For example, you should be able to make a quick schematic sketch of a neuron (nerve cell), with the major parts correctly labeled.

A Respect for Ethics and for People

You can count on your psychology instructor's having one more expectation of you. Psychology is a science, but, as a *life* science, it has a strong ethical dimension. Most psychology instructors are scientists who enter the field with a profound sense of ethics and a dedication to improving the quality of life. Cultivating a similar attitude will not only enhance your appreciation and enjoyment of the course, but will put you more closely in tune with the instructor's mind-set. And that is never a bad thing.

Students should learn that human behavior is guided by principles that can be studied empirically, just as we study the physical world. As this understanding develops, they should begin to distinguish between information that results from a scientific process and pop psychology. They learn that intuition is not a reliable source of information about human behavior. They also learn that individual studies are limited and can lead to incorrect or incomplete conclusions, but that the scientific process prevails over intuition because it progresses toward greater understanding when converging evidence is the basis for decisions. Within this overarching framework, students need to develop a working knowledge of the vocabulary and core phenomena in psychology, focusing in foundation courses on information that is useful for being an informed consumer of scientific knowledge across disciplines. Because students often begin college without an appreciation of how information is organized into concepts, the presentation and testing of material should lead them to see information as organized into general principles and supporting observations, and they should be able to discuss the relationships between concepts and specific observations.

Students often come to college with very little experience learning on their own. They are accustomed to teachers who explicitly transmit every key term, and many students learn from texts predominantly by memorizing facts as unrelated pieces of information. In college, however, the quantity of material is too great to continue this strategy. Students need to learn to monitor their own understanding of material by engaging in a dialogue with themselves about how information is organized into concepts, and how those concepts relate to what they already know. The most useful first step is to learn to approach a text in a new way. I teach them to scan the chapter outline, then turn immediately to the summary, read it, and turn to another subject for that evening. This gives them an overview of what the major concepts in the chapter are and allows them time to think about those concepts without interference from reading related information. I ask them next to take several evenings to read the chapter, taking care to read the comics, illustrations, and special topic boxes that help them understand and remember the information. After reading the summary again, they need to master the critical terms by engaging in an internal dialogue about those terms. They might, for example, write and answer their own essay question about classical conditioning that requires them to label the major terms involved in that form of learning, or they might rehearse the neuron by drawing and labeling a diagram. They need to realize that study guides are generally ineffective for initial learning of material, but that they are very effective to help students identify gaps in their knowledge after they believe that they have mastered the material.

—Debra Ann Poole,
Professor, Central Michigan University

CHAPTER 2

KEYS TO SUCCESSFUL STUDY

THIS CHAPTER OUTLINES SKILLS THAT ARE INDISPENSABLE TO SUCCESSFUL STUDY, with special emphasis on skills important to the study of psychology on the introductory level.

But let's not start thinking about psychology just yet. Let's just start thinking. After all, isn't that what college and college courses are all about?

Well, not quite. *Think* about it.

For the ancient Greeks of Plato's day, about 428 to 348 B.C., "higher education" really *was* all about thinking. Through dialogue, back and forth, the teacher and his student *thought* about psychology, mathematics, physics, the nature of reality—whatever. Perhaps the teacher evaluated the quality of his student's thought, but there is no evidence that Plato graded exams, let alone assigned the student a final grade for the course.

B.C. was a long time ago. Times have changed.

"Don't Study for the Grade!"

Today, you get graded. All the time, and on everything you do. Now, most of the professors, instructors, and teaching assistants from whom you take your courses will tell you that the "real value" of the course is in its contribution to your "liberal education." Professors may even solemnly protest that they do *not* "teach for" the midterm and final. Nevertheless, the introductory, survey, and core courses almost always include at least one midterm and a final, and these almost always account for a very large part of the course grade. Even if the professor can disclaim with a straight face *teaching for* these exams, few students would deny *learning for* them.

The truth is this: more than anything else, most curriculum core courses *are* focused on the midterm and final exams.

"Grades Aren't Important"

Let's keep thinking.

You and, most likely, your family are investing a great deal of time and cash and sweat in your college education. It *would* be pretty silly if the payoff of all those resources was a letter grade and a numerical GPA. Ultimately, of course, the payoff is knowledge, a feeling of achievement, an intellectually and spiritually enriched life—*and* preparation for a satisfying and (you probably hope) financially rewarding career.

But the fact is that if you don't perform well on midterms and finals, your path to all these forms of enrichment will be blocked. And the fact *also* is that your performance is measured by grades. Sure, almost any reason you can think of for investing in college is more important than amassing a collection of A's and B's, but those stupid little letters are part of what it takes to get you to those other, far more important, goals.

"Don't Study *for the* Exam"

Most professors hate exams and hate grades. They believe that prodding students to pass tests and then evaluating their performance with a number or letter makes the whole process of education seem pretty trivial. Those professors who tell you that they "don't teach for the exam" may also advise you not to "study for the exam." That's not exactly what they mean. They want you to study, but to study in order to learn, not *for* the exam.

It's well-meaning advice, and it's true that if you study *for* the exam, intending to ace it, then promptly forget everything you've "learned," you are making a pretty bad mistake. Yet those same professors are part of a system that demands exams and grades, and if you don't study for and *for* the exam, the chances are very good that you won't make the grade and you won't achieve the higher goals you, your family, *and your professors* want for you.

Lose-Lose or Win-Win? Your Choice

When it comes to studying, especially in your introductory-level courses, you have some choices to make. You can decide grades are stupid, not study, and perform poorly on the exams. You can try not to study for the exams, but concentrate on higher goals, perform poorly on the exams, and never have the opportunity to reach those higher goals. You can study *for* the exams, ace them all, then flush the information from your memory bank, collect your A or B for the course, and move on without having learned a thing. These are all lose-lose scenarios, in which no one—neither you nor your teacher (nor your family, for that matter)— gets what he, she, or they really want.

Or you can go the win-win route.

We've used the words *goal* and *goals* several times. For an army general, winning the war is the goal, but to achieve this goal the general must first accomplish certain *objectives*, such as winning battle number one, number two, number three, and so on. Objectives are intermediate goals or steps toward an ultimate goal.

Now, put exams in perspective. Performing well on an exam need not be an alternative to achieving higher goals, but should be an objective necessary to achieving those higher goals.

The win-win scenario goes like this: Use the fact of the exams as a way of focusing your study for the course. Focus on the exams as immediate objectives, crucial to achieving your ultimate goals. *Do* study for the exam, but not *for* the exam. *Don't* mistake the battle for the war, the objective for the goal; but *do* realize that you must attain the objectives in order to achieve the goal.

And *that*, ladies and gentlemen, is the purpose of this book:

- To help you ace the midterm and final exams in introductory-level psychology courses…
- Without forgetting everything you learned after you've aced them.

This guide will help you *use* the exams to master the course material. This guide will help you make the grade—*and* actually learn something in the process.

Focus

How many times have you read a book, word for word, finished it, and closed the cover—only to realize that you've learned almost nothing from it? Unfortunately, it's something we all experience. It's not that the material is too difficult or that it's over our heads. It's that we mistake reading for studying.

For many of us, reading is a passive process. We scan page after page, the words go in, and, alas, the words seem to go out. The time, of course, goes by. We've *read* the book, but we've *retained* all too little.

Studying certainly involves reading, but reading and studying aren't one and the same activity. Or we might put it this way: studying is intensely focused reading.

How do you *focus* reading?

Begin by setting objectives. Now, saying that your objective in reading a certain number of chapters in a textbook or reading your lecture notes is to "learn the material" is not very useful. It is an obvious but vague *goal* rather than a well-defined *objective.*

Why not let the approaching exam determine your objective?

"I will read and retain the stuff in Chapters 10 through 20 because that's what's going to be on the exam, and I want to ace the exam."

Now you at least have an objective. Accomplish this *objective,* and you will be on your way to achieving the *goal* of "learning the material."

◆ An *objective* is an immediate target. A *goal* is for the long term.

Concentrate

To move from passive reading to active study requires, first of all, concentration. Setting up objectives (immediate targets) rather than looking toward goals (long-term targets) makes it much easier for you to concentrate. Few of us can (or would) put our personal lives on hold for four years of college, several years of graduate school, and *X* years in the working world in order to concentrate on achieving a *goal.* But just about anyone can discipline him- or herself to set aside distractions for the time it takes to achieve the *objective* of studying for an exam.

Step 1. Find a quiet place to work.

Step 2. Clear your mind. Push everything aside for the few hours you spend each day studying.

Step 3. Don't daydream—*now.* Daydreaming, letting your imagination wander, is actually essential to real learning. But, right now, you have a specific objective to attain. This is not the time for daydreaming.

Step 4. Deal with your worries. Those pressing matters you can do something about *right now,* take care of. Those you can't do anything about, push aside for now.

Step 5. Don't *worry* about the exam. Take the exam seriously, but don't fret. Instead of worrying about the prospect of failure, use your time to eliminate failure as an option.

Plan

Let's go back to that general who knows the difference between objectives and goals. Chances are he or she also knows that you'd better not march off to battle without a plan. Remember, you don't want to *read.* You want to *study.* This requires focusing your work with a plan.

The first item to plan is your time.

Step 1. Dedicate a notebook or organizer-type date book to the purpose of recording your scheduling information.

Step 2. Record the following:
a. Class times
b. Assignment due dates
c. Exam dates
d. Extracurricular commitments
e. Study time

Step 3. Inventory your various tasks. What do you have to do today, this week, this month, this semester?

Step 4. Prioritize your tasks. Everybody seems to be grading you. Now's *your* chance to grade the things they give you to do. Label high-priority tasks "A," middle-priority tasks "B," and lower-priority tasks "C." This will not only help you decide which things to do first, it will also aid you in deciding how much time to allot to each task.

Step 5. Enter your tasks in your scheduling notebook and assign order and duration to each according to its priority.

TIP: If you are in doubt about what tasks to assign the highest priority, it is generally best to allot the most time to the most complex and difficult tasks and to get these done first.

Step 6. Check off items as you complete them.

Step 7. Keep your scheduling book up to date. Reschedule whatever you do not complete.

Step 8. Don't be passive. Actively *monitor* your progress toward your objectives.

Step 9. Don't be passive. Arrange and rearrange your schedule to get the most time when you need it most.

Packing Your Time

Once you have found as much time as you can, pack it as tightly as you can.

Step 1. Assemble your study materials. Be sure you have all necessary textbooks and notes on hand. If you need access to library reference materials, study in the library. If you need access to reference materials on the Internet, make sure you're at a computer.

Step 2. Eliminate or reduce distractions.

Step 3. Become an efficient reader and note taker.

An Efficient Reader

Step 3 requires further discussion. Let's begin with the way you read.

Nothing has greater impact on the effectiveness of your studying than the speed and comprehension with which you read. If this statement prompts you to throw up your hands and wail, "I'm just not a fast reader," don't despair. You can learn to read faster and more efficiently.

Consider taking a speed-reading course. Take one that your university offers or endorses. Most of the techniques taught in the major reading programs actually do work. Alternatively, do it yourself.

Step 1. When you sit down to read, try consciously to force your eyes to move across the page faster than normal.

Step 2. Always keep your eyes moving. Don't linger on any word.

Step 3. Take in as many words at a time as possible. Most slow readers aren't slow-witted. They've just been taught to read word by word. Fast and efficient readers, in contrast, have learned to read by taking in groups of words. Practice taking in blocks of words.

Step 4. Build on your skills. Each day, push yourself a little harder. Move your eyes across the page faster. Take in more words with each glance.

Step 5. Resist the strong temptation to fall back into your old habits. Keep pushing.

TIP: Are you a—*vocalizer?* A vocalizer is a reader who, during silent reading, either mouths the words or says them mentally. Vocalizing greatly slows reading, often reduces comprehension, and is just plain tiring. Work to overcome this habit—*except* when you are trying to memorize some specific piece of information. Many people do find it helpful to say over a sentence or two in order to memorize its content. Just bear in mind that this does not work for more than a sentence or paragraph of material.

When you review material, consider skimming rather than reading. Hit the high points, lingering at places that give you trouble and skipping over the stuff you already know cold.

An Interactive Reader

Early in this chapter we contrasted passive reading with active studying. A highly effective way to make the leap from passivity to activity is to become an *interactive* reader.

Step 1. Read with a pencil in your hand.

Step 2. Use your pencil to underline key concepts. Do this consistently. (That is, *always* read with a pencil in your hand.) Don't waste your time with a ruler; underscore freehand.

Step 3. Underline *only* the key concepts. If everything seems important to you, then up the ante and underline only the absolutely *most* important passages.

TIP: The physical act of underlining actually helps you memorize material more effectively—though no one is quite sure why. Furthermore, underlining makes review skimming more efficient and effective.

Step 4. If you prefer to highlight material with a transparent marker, fine. But you'll still need a pencil or pen nearby. Carry on a dialogue with your books by writing condensed notes in the margin.

Step 5. Put difficult concepts into your own words—right in the margin of the book. This is a great aid to understanding and memorization.

Step 6. Link one concept to another. If you read something that makes you think of something else related to it, make a note. The connection is almost certainly a valuable one.

Step 7. Comment on what you read.

TIP: Some instructors advise against using highlighter-style markers to underscore books and notes, because doing so may discourage you from writing notes in the margin. If holding a highlighter means that you won't also pick up a pencil to engage in a lively dialogue with books or notes, then it *is* best to lay aside the highlighter and take the more *actively* interactive route.

Taking Notes

The techniques of underlining, highlighting, paraphrasing, linking, and commenting on textbook material can also be applied to your classroom and lecture notes.

Of course, this assumes that you have taken notes. There are some students who claim that it is easier for them to listen to a lecture if they *avoid* taking notes. For a small minority, this may be true; but the vast majority of students find that note taking is essential when it comes time to study for midterms and finals. This does not mean that you should be a stenographer or court reporter, taking down each and every word. To the extent that it is possible for you to do so, absorb the lecture *in your mind*, then jot down major points, preferably in loose outline form.

TIP: Some students are reluctant to write in their textbooks because it reduces resale value. True enough. But is it worth an extra $5 at the end of the semester if you don't get the most out of your multi-thousand-dollar and multi-hundred-hour investment in the course?

Become sensitive to the major points of the lecture. Some lecturers will come right out and tell you, "The following three points are key." That's your cue to write them down. Other cues include:

- **Repetition.** If the lecturer repeats a point, write it down.
- **Excitement.** If the lecturer's voice picks up, if his or her face becomes suddenly animated, if, in other words, it is apparent that the material is at this point of particular interest to the speaker, your pencil should be in motion.
- **Verbal cues.** In addition to such verbal elbows in the rib as "this is important," most lecturers underscore transitions from one topic to another with phrases such as "Moving on to . . ." or "Next, we will . . ." or the like. This is your signal to write a new heading.

TIP: Note taking is especially important in fields such as psychology, because new studies are continually being published, many of which relate to fundamental issues relevant even in introductory courses. Lectures are up to date, while your textbook is obsolete the moment it is printed.

- ***Slowing down.*** If the lecturer gives deliberate verbal weight to a word, phrase, or passage, make a note of it.
- **Visual aids.** If the lecturer writes something on a blackboard or overhead projector or in a computer-generated presentation, make a note.

TIP: No one can tell you just how much to write, but bear this in mind: most lecturers read from notes rather than fully composed scripts. Four double-spaced type-written or word-processed pages (about a thousand words *in note form*) represent sufficient note material for an hour-long lecture. Ideally, you might aim at producing about 75 percent of this word count in the notes you take—perhaps 750 words during an hour-long lecture.

Filtering Notes

Some students take neat notes in outline form. Others take sprawling, scrawling notes that are almost impossible to read. Most students can profit from *filtering* the notes they take. Usually, this does *not* mean rewriting or retyping your notes. Many instructors agree that this is a waste of time. Instead, they advise, underscore the most important points, filtering out the excess.

If you have taken notes on a notebook or laptop computer, consider arranging the notes in clear outline form. If you have handwritten notes, however, it may not be worth the time it takes to create a neat outline. Instead, spend that time merely underlining or highlighting the most important concepts.

TIP: While many instructors do not recommend rewriting or typing up your lecture notes, this may be useful in a subject like psychology, which requires memorizing a good deal of technical information and specialized vocabulary. You may want to give this tactic a try. If it helps, use it. However, if you find that rewriting or typing your notes does not help you, drop the practice.

Tape It?

Should you bring a tape recorder to class? The short answer is, probably not. To begin with, some instructors object to having their lectures recorded. Even more important, however, is the tendency to complacency that a tape recorder creates. You might feel that you don't have to listen very carefully to the live lecture, since you're getting it all on tape. This could be a mistake, since the live presentation is bound to make a greater impression on you, your mind, and your memory than a recorded replay.

TIP: Before you tote your laptop or notebook computer to class, make certain that the instructor approves of such note-taking devices in the classroom. Most lecturers have no problem with these, but some find the tap-tap-tapping of maybe more than a hundred students distracting.

So much for the majority view on tape recorders; however, one contributor to this book actually *recommends* to his students that they tape lectures. He even suggests that they avoid taking written notes in class and instead play back the tape later and take notes from that. If this method appeals to you, make sure of the following:

- Clear with your instructor the use of a tape recorder. Make sure he or she has no objections.
- Before you rely on the tape recorder, make sure that it works, that its microphone adequately picks up what's going on in the classroom, and that you are

well positioned to record. The back of a large lecture hall is probably not a good place to be.

- Don't "zone out" or daydream during the lecture, figuring that you don't need to pay attention because you're getting it all on tape anyway. The value of the tape-recorded lecture is as a repetition and reinforcement of what you have already absorbed.
- Make certain that you really do play back—and take notes from!—the taped lectures. This means setting aside the time to repeat the entire lecture.

Remember: The mere fact that you have tape-recorded a lecture does not mean that you have learned the material. Technology is indeed marvelous, but it won't perform miracles. You have to listen to the tape and take notes from it.

Build Your Memory

Just as a variety of speed-reading courses are available, so a number of memory-improvement courses, audiotapes, and books are on the market. It might be worth your while to scope some of these out, especially for a memory-intensive course like introductory psychology or if you are planning ultimately to go into a field that requires the memorization of a lot of facts. In the meantime, here are some suggestions for building your memory:

TIP: Memorization is important, make no mistake, especially in an information-rich field like psychology. However, it *is* usually overrated. Virtually all of the instructors and professors who have contributed to this book counsel students to *think* rather than merely memorize. This is true even in introductory courses that rely mainly on multiple-choice exams. Most psychology instructors who use multiple-choice exams make an effort to create multiple-choice questions that test broad and basic concepts as well as factual material. Many professors, even in introductory courses, also include short-answer and essay questions in exams. These are specifically aimed at evaluating how students handle ideas and concepts, not just rote-learned facts. One contributor to this books uses the essay-exam format exclusively.

- Be aware that most so-called memory problems are really learning problems. You "forget" things you never *learned* in the first place. This is usually a result of passive reading or passive attendance at lectures—the familiar in-one-ear-and-out-the-other syndrome.
- Memorization is made easier by two seemingly opposite processes. First, we tend to remember information for which we have a context. It will indeed be hard to remember a bunch of definitions if you try to study these out of context—as a list rather than as parts of an interrelated system of thought.
- Second, memorization is often made easier if we break down complex material into a set of key phrases or words. You may find it easier to memorize the key points relating to inflation and unemployment than a narrative description of these two phenomena.
- It follows, then, that the best way to build your memory where a certain subject is concerned is to try to understand information in context. Get the "big picture."
- It also follows that, even if you have the big picture, you may want to break down key concepts into a few key words or phrases.

How We Forget

It is always better to keep up with class work and study than it is to fall behind and desperately struggle to catch up. This said, it is nevertheless true that most forgetting occurs within the first few days of exposure to new material. That is, if you learn 100 facts about Subject A on December 1, you may forget 20 of those facts by December 5 and another 10 by December 10, but by March 1 you may still remember 50 facts. The curve of your forgetting tends to flatten. Eventually, you may forget all 100 facts, but you will forget fewer and fewer each week.

Now, what does this mean to you?

It means that, midway through the course, you need to review material you learned earlier in the course. You cannot depend on having mastered it forever by having studied it two, three, four, or more weeks earlier.

TIP: Many memory experts suggest that you try to put the key terms you identify into some sort of sentence; then memorize the sentence. Others suggest creating an acronym out of the initials of the key words or concepts. No one who lived through the Watergate scandal in the 1970s can forget that President Nixon's political campaign was run by CREEP (Committee to *RE*-Elect the President), and long after everything else taught in high school biology is forgotten, many students remember the sentence they memorized in order to learn how biologists classify organisms: *Ken Put Candy On Fred's Green Sofa* (kingdom, phylum, class, order, family, genus, species). Use whatever memory aids help *you*.

The Virtues of Cramming

Ask any college instructor about last-minute cramming for an exam, and you'll almost certainly get a knee-jerk condemnation of the practice. But maybe it's time to think beyond that knee jerk.

Let's get one thing absolutely straight. You cannot expect to pack a semester's worth of studying into a single all-nighter. It just isn't going to happen. However, cramming can be a valuable *supplement* to a semester of conscientious studying.

- You forget the most within the first few days of studying. (Or have you forgotten?)

Well, if you cram the night before the exam, those "first few days" won't fall between your studying and the exam, will they?

- Cramming creates a sense of urgency. It brings you face to face and toe to toe with your objective. Urgency concentrates the mind.
- Assuming you aren't totally exhausted, material you study within a few hours of going to bed at night is more readily retained than material studied earlier in the day.

Burning the midnight oil may not be such a bad idea.

TIP: Any full-time college student studies several subjects each semester. This makes you vulnerable to interference—the possibility that learning material from one subject will interfere with learning material in another subject. Interference is usually at its worst when you are studying two similar or related subjects. If possible, arrange your study time so that work on similar subjects is separated by work on an unrelated one.

Cramming Cautions

Then again, staying up late before a big exam may not be such a hot idea, either. Don't do it if you have an early-morning exam. And don't transform cramming into an all-nighter. You almost certainly need some sleep to perform competently on tomorrow's exam.

Remember, too, that while cramming creates a sense of urgency, which may stimulate and energize your study efforts, it may also create a feeling of panic, and panic is never helpful.

Cramming is *not* a substitute for diligent study throughout the semester. But just because you *have* studied diligently, don't shun cramming as a *supplement* to regular study, a valuable means of refreshing the mind and memory.

TIP: If you *hate* cramming, don't do it. It's not for you, and it will probably only raise your anxiety level. Get some sleep instead.

Polly Want a Cracker?

We've been talking a lot about memory and memorization. It's an important subject and, for just about any course of study, an important skill. Some subjects—psychology included—are more fact-and-memory intensive than others. However, beware of relying too much on simple, brute memory. Try to assess what the professor really wants: students who demonstrate on exams that they have absorbed the facts he or she and the textbooks have dished out? Or students who demonstrate such skills as critical thinking, synthesis, analysis, and imagination? Depending on the professor's personal style and the kind of exam he or she gives (predominantly essay versus predominantly multiple choice, for example), you may actually be penalized for parroting lectures. ("I know what *I* think. I want to know what *you* think.")

Use *This* Book (and Get Old Exams)

One way to judge what the professor values and expects is to pay careful attention in class. Is discussion invited? Or does the course go by the book and by the lecture? Also valuable are previous exams. Many professors keep these on file and allow students to browse through them freely. Fraternities, sororities, and formal as well as informal study groups sometimes maintain such files, too. These days, previous exams may even be posted on the university department's World Wide Web site.

Of course, you are holding in your hand a book chock-full of sample and model midterm and final exams. Read them. Study them. And let them focus your study and review of the course.

Study Groups: The Pro and the Con

In the old days (whenever that was), it was believed that teachers *taught* and students *learned*. More recently, educators have begun to wonder whether it is possible to teach at all. A student, they say, *learns* by teaching him- or herself. The so-called teacher (who might better be called a "learning facilitator") helps the student teach him- or herself.

Well, maybe this is all a matter of semantics. Is there really a difference between *teaching* and *facilitating learning*? And between *learning* and *teaching oneself*? The more important point is that the focus in education has turned away from the teacher to the student, and students, in turn, have often responded by organizing study groups, in which they help each other study and learn.

These can be very useful:

- In the so-called real world (that is, the world after college), most problems are solved by teams rather than individuals.
- Many people come to an understanding of a subject through dialogue and question and answer.
- Studying in a group (or even with one partner) makes it possible to drill and quiz one another.
- In a group, complex subjects can be broken up and divided among members of the group. Each one becomes a specialist on some aspect of the subject, then shares his or her knowledge with the others.
- Studying in a group may improve concentration.
- Studying in a group may reduce anxiety.

Not that study groups are without their problems:

- All too often, study groups become social gatherings, full of distraction (however pleasant) rather than study. This is the greatest pitfall of a study group.
- All members of the group must be highly motivated to study. If not, the group will become a distraction rather than an aid to study, and it is also likely that friction will develop among the members, some of whom will feel burdened by "freeloaders."
- The members of the study group must not only be committed to study, but to one another. Study groups fall apart—bitterly—if members, out of a sense of competition, begin to withhold information from one another. This *must* be a Three Musketeers deal—all for one and one for all—or it is worse than useless.
- The group may promote excellence—or it may agree on mediocrity. If the latter occurs, the group will become destructive.

In summary, study groups tend to bring out the members' best as well as worst study habits. It takes individual and collective discipline to remain focused on the task at hand, to remain committed and helpful to one another, to insist that everyone shoulder his or her fair share, and to insist on excellence of achievement as the only acceptable standard—or, at least, the only valid reason for continuing the study group.

CHAPTER 3

SECRETS OF SUCCESSFUL TEST TAKING

SOMETIMES IT SEEMS THAT THE DIFFERENCE BETWEEN ACADEMIC SUCCESS AND something less than success is not smarts versus nonsmarts or even study versus nonstudy, but simply whether or not a person is "good at taking tests." That phrase—"good at taking tests"—was probably first heard back when the University of Bologna opened for business late in the eleventh century. The problem with phrases like this is that while they are true enough, they are not very helpful.

Fact: Some people *are* and some people *are not* good at taking tests.

So what? Even if successful test taking doesn't come easily or naturally to you, you *can* improve your test taking skills. Now, if you happen to have a knack for taking tests, well, congratulations! But that won't help you much if you neglect the kind of preparation discussed in the previous chapter.

Why Failure?

In analyzing performance on most tasks, it is generally better to begin by asking what you can do to succeed. But in the case of taking tests, success is largely a matter of avoiding failure. So let's begin there.

When the celebrated bank robber Willy "the Actor" Sutton was caught, a reporter asked him why he robbed banks. "Because that's where the money is," the handcuffed thief replied. At least one answer to the question of why some students perform poorly on exams is just as simple: "Because they don't know the answers."

There is no magic bullet in test taking. But the closest thing to one is *knowing the course material cold.* Pay attention, keep up with reading and other assignments, attend class, listen in class, take effective notes—in short, follow the recommendations of the previous chapter—and you will have taken the most important step toward exam success.

Yet have you ever gotten your graded exam back, with disappointing results, read it over, and found question after question that you realize you *could* have answered correctly?

"I *knew* that!" you exclaim, smacking yourself in the forehead.

What happened? You really *did* prepare. You really *did* know the material. What happened?

TIP: Most midterm and final exams really *are* representative of the course. If you have mastered the course material, you will almost certainly be prepared to perform well on the examination. Very few instructors purposely create deceptive exams or trick questions or even questions that require you to think beyond the course. Most instructors are interested in creating exams that help you and them evaluate your level of understanding of the course material. Think of the exams as natural, logical features of the course, not as sadistic assignments designed to trip you up. Remember, your success on an examination is also a measure of the instructor's success in presenting complex information. Very few teachers can—or want to—build careers on trying to *fail* their students.

Anxiety, Good and Bad

The great American philosopher and psychologist William James (1842–1910) once advised his Harvard students that an "ounce of good nervous tone in an examination is worth many pounds of . . . study." By "good nervous tone" James meant something very like anxiety. You *should not* expect to feel relaxed just before or during an exam. Anxiety is natural.

Anxiety is natural because it is helpful. "Good nervous tone," alert senses, sharpened perception, and adrenaline-fueled readiness for action are *natural* and *healthy* responses to demanding or threatening situations. We are animals, and these are reactions we share with other animals. The mongoose that relaxes when it confronts a cobra is a dead mongoose. The student who takes it easy during the psych final . . . Well, the point is neither to fight anxiety nor to fear it. Accept it, and even welcome it as an ally. Unlike our hominid ancestors of distant prehistory, we no longer need the biological equipment of anxiety to help us fight or fly from the snapping saber teeth of some animal of prey, but every day we do face challenges to our success. Midterms and finals are just such challenges, and the anxiety they provoke is real, natural, and unavoidable. It may even help us excel.

What good can anxiety do?

- Anxiety can focus our concentration. It can keep the mind from wandering. This makes thought easier, faster, and, often, more acute and effective.
- Anxiety can energize us. We've all heard stories about a 105-pound mother who is able to lift the wreckage of an automobile to free her trapped child. This isn't fantasy. It really happens. And just as adrenaline can provide the strength we need when we need it most, it can enhance our ability to think under pressure.

TIP: Exam questions are battles in a war. No general expects to win every battle. Accept your losses and move on. Dwell on your losses, and you will continue to lose.

- Anxiety moves us along. We work faster than when we are relaxed. This is valuable, since, for most midterms and finals, limited time is part of the test.
- Anxiety prompts us to take risks. We've all been in classes in which the instructor has a terrible time trying to get students to speak and discuss and venture an opinion. "Come on, come on," the poor prof protests. "This is like pulling teeth!" Yet, when exam time comes, all heads are bent over blue books or scantron answer sheets, and the answers—*some* kind of answers—are flowing forth or, at least, grinding out. Why? Because the anxiety of the exam situation overpowers the inertia that keeps most of us silent most of the time. We take the risks we have to take. We answer the questions.
- Anxiety can make us more creative. This is related to risk taking. "Necessity," the old saying goes, "is the mother of invention." Phrased another way; *we do what we have to do.* Under pressure, many students find themselves taking fresh and creative approaches to problems.

TIP: More serious stimulant drugs (speed, uppers) are *never* a good idea. They are both illegal and dangerous (possibly even deadly), and, for that matter, their effect on exam performance is unpredictable. The chances are that they will impede your performance rather than aid it (although you may erroneously *feel* that you are doing well). Also avoid over-the-counter stimulants. These are caffeine pills and will probably increase your anxiety rather than improve your performance.

So don't shun anxiety. But, unfortunately, the scoop on anxiety isn't all good news, either.

Anxiety evolved as a mechanism of *physical* survival. Biologists and psychologists talk of the fight-or-flight response. Anxiety prepares a threatened animal either to fight the threat or to flee from it. The action is physical and, typically, very short term. In our "civilized" age, however, the threats are generally less physical than intellectual and emotional, and they tend to be of longer duration than a physical fight or a physical flight. This means that the anxiety mechanism does not always work to enhance our chances for "survival" or, at least, our chances to survive the course by performing well on exams. Some of us are better than others at adapting the *physical* benefits of anxiety to the *intellectual* and *emotional* challenges of an exam. Some of us, unfortunately, are unable to benefit from anxiety, and for still others of us, anxiety is downright harmful. Here are some of the negative effects anxiety may have on exam performance:

- Anxiety can make it difficult to concentrate. True, anxiety focuses concentration. But if it focuses concentration on the anxious feelings themselves, you will have less focus left over for the exam. Similarly, anxiety may cause you to focus unduly on the perceived consequences of failure.
- Anxiety causes carelessness. If anxiety can prompt you to take creative risks, it can also cause you to rush through material and, therefore, to make careless

mistakes or simply to fail to think through a problem or question.

- Anxiety distorts focus. Anxiety may impede your judgment, causing you to give disproportionate weight to relatively unimportant matters. For example, you may become fixated on solving a lesser problem at the expense of a more important one. This is related to the next point.
- Anxiety may distort your perception of time. You may think you have more or less of it than you really do. The result may be too much time spent on a minor question at the expense of a major one.
- Anxiety tends to be cumulative. Many test takers have trouble with a question early in the exam, then devote the rest of the exam to worrying about it instead of concentrating on the *rest of the exam.*
- Anxiety drains energy. For short periods of time, anxiety can be energizing and invigorating. But if anxiety becomes chronic, it begins to tire you out. You do not perform as well.
- Anxiety can keep you from getting the rest you need. If it is generally unwise to stay up all night *studying* for an exam, how much less wise is it to stay up uselessly *worrying* about one?

TIP: Don't make the mistake of devoting all your time to trying to make *last-minute* repairs to weak spots ("I've got one hour to read that textbook I should have been reading all along!") and ignoring your strengths. Develop your strengths. With any luck at all, the exam will give you an opportunity to show yourself at your best—not just trip you up at your worst. Be as prepared as you can be, but, remember, there is nothing wrong with excelling in a particular area. Play to your strengths, not your weaknesses.

How can you combat anxiety?

Step 1. *Don't* fight it. Accept it. Remember, anxiety is a *natural* response to a stressful situation. Remember, too, that some degree of anxiety aids performance. Try to learn to accept anxiety and *use* it. Let it sharpen your wits and stoke the fires of your creativity.

Step 2. Don't worry about how you feel. Focus on the task. Usually, you will feel better once you overcome the initial jitters and inertia. William James, who lauded "good nervous tone," also once observed that we do not run because we are frightened, but we are frightened because we run. If you concentrate on your fear and act as if you are afraid, you will become even more fearful.

Step 3. Prepare for the exam. Do whatever you must do to master the material. Build confidence in your understanding of the course, and your anxiety should be reduced.

Step 4. Get a good night's sleep before the exam.

Step 5. Avoid coffee and other stimulants. Caffeine tends to increase anxiety. (However, if you are a caffeine fiend, don't pick the day or two before a big exam to kick the habit. You *will* suffer withdrawal symptoms.)

Step 6. Try to get fresh air shortly before the exam. This is especially valuable if

TIP: When you study for an exam, it is usually best to assign high priority to the most complex and difficult issues, devote ample time to these, and master them first. When you *take* an exam, however, and you are under time pressure, tackle first what you can most readily and thoroughly answer, then go on to more doubtful tasks. Your professor will be more favorably impressed by good or correct answers than by failed attempts to answer questions you find difficult.

you have been cooped up for a long period of study. Take a walk. Get a look at the wider world for a few minutes.

Have a Plan, Make a Plan

A large component of destructive anxiety—probably the largest—is fear of the unknown. Reduce anxiety by taking steps to reduce the component of the unknown.

Step 1. To repeat—do whatever is necessary to master the material on which you will be examined.

Step 2. Use the exams in this book to familiarize yourself with the kinds of exams you are likely to encounter.

Step 3. If possible, examine old exams actually given in the course.

Let's pause here before going on to Step 4. Just reading over the exams in this book or leafing through exams previously given in the course will not help you much. Analyze:

- The types of questions asked. Are they essay or objective questions? (We'll discuss these shortly.) Do they call for "regurgitation" of memorized material, or are they more "think"-oriented, requiring significant initiative to answer?
- Don't just predict which questions you could and could not answer. Try actually answering some of the questions.
- If you are looking at sample exams with answers, evaluate the answers. How would you grade them? What would you do better?
- Don't just sit there, *do* something. If your analysis of the sample exams or old exams reveals areas in which you are weak, address those weaknesses.

An effective way to reduce the unknown is to create a plan for confronting it. Let's go on to Step 4.

Step 4. Make a plan. Begin *before* the exam. Decide what areas you need to study hardest. Based on your textbook notes and—especially—on your lecture notes, try to anticipate what kinds of questions will be asked. Work up answers or sketches of answers for these.

Step 5. Make sure you've done the simple things. The night before your exam, make certain that you have whatever equipment you'll need. If you will be allowed to use reference materials, bring them. If you are permitted to write on a laptop or notebook computer, make certain your batteries are fully charged. If you are writing the exam longhand, make certain you have pens, pencils, paper. Bring a watch.

Step 6. Expect a shock. The first sight of the exam usually packs a jolt. At first sight, questions may draw a blank from you. Questions you were sure would appear on the exam will be absent, and some you never expected will be staring you in the face. Don't panic. *Everybody feels this way.*

Step 7. Write nothing yet. Read through the exam thoroughly. In the case of lengthy multiple-choice exams, skim over all of the questions. Be certain that you (1) understand any instructions and (2) understand the questions.

Step 8. If you are given a choice of which questions to answer, choose them now. Unless the questions vary in point value assigned to them, choose those that you feel most confident about answering. Don't challenge yourself.

Alternative Step 8: If you are required to answer all the questions, identify those about which you feel most confident. Answer these first.

Step 9. After you have surveyed the exam, create a time budget: note—jot down—how much time you should give to each major question.

Step 10. Reread the question before you begin to write. Then plan your answer.

Plan Your Answer

Perhaps you have heard a teacher or professor comment on the exam he or she has just handed out: "The answers are in the questions." This kind of remark is more helpful than it may at first seem.

Begin by looking for the key words in the question. These are the verbs that *tell* you what to do, and they typically include:

- Compare
- Contrast
- Criticize
- Define
- Describe
- Discuss
- Evaluate
- Explain
- Illustrate
- Interpret
- Justify
- Outline
- Relate

- Review
- State
- Summarize
- Trace

You will find most of these key words in essay questions rather than in short-answer, multiple-choice, or fill-in-the-blank questions, so we will have much more to say about the key words in Chapter 5, which is devoted to answering the essay exam. But be aware that the following key words are often found even in quite brief short-answer questions:

- To ***define*** something is to state the precise meaning of the word, phrase, or concept. Be succinct and clear.
- To ***illustrate*** is to provide a specific, concrete example.
- To ***outline*** is to provide the main features or general principles of a subject. This need not be in paragraph or essay form. Often, outline format is expected.
- To ***state*** is similar to ***define,*** though a statement may be even briefer and usually involves delivering up something that has been committed to memory.
- To ***summarize*** is briefly to state—in sentence form— the major points of an argument or position or concept, the principal features of an event, or the main events of a period.

As has just been mentioned, advice on answering essay examinations is the subject of Chapter 5. For the moment, just be aware that you will want to budget time for creating a scratch outline of your essay answer.

Approaching the Short-Answer Test

While the value of planning is more or less obvious in the case of essay exams, you will also find it valuable in objective, multiple-choice, or other short-answer exams. These are of two major kinds:

1. *Recall exams* include questions that call for a single short answer (usually there is a single "correct" answer) and fill-in-the-blank questions, in which you are asked to supply missing information in a statement or sentence.
2. *Recognition exams* include multiple-choice tests, true-or-false tests, and matching tests (match one from column A with one from column B).

If the exam is a long one and time is short, invest a few minutes in surveying the questions, so that you can be certain to answer those you are confident of, even if they come near the end of the exam.

Be prepared to answer multiple-choice questions through a process of elimination, if necessary. Usually, even if you are uncertain of the one correct answer among a choice of five, you *will* be able to eliminate one, two, or three answers you know are *incorrect.* This at least increases your odds of giving a correct response.

Unless your instructor has specifically informed you that he or she is penalizing guesses (actually taking points away for incorrect responses versus awarding zero points to unanswered questions), *do* guess the answers even to those questions that leave you in the dark.

Plan your responses to true-or-false questions carefully. Look for telltale qualifying words, such as *all, always, never, no, none,* or *in all cases.* Questions with such absolute qualifiers *often* require an answer of *false,* since relatively few general statements are always either true or false. Conversely, questions containing such qualifiers as *sometimes, usually, often,* and the like, are frequently answered correctly with a response of *true.*

A final word on guessing: First guess, best guess. Statistical evidence consistently shows that a first guess is more likely to be right than a later one. Obviously, if you have responded one way to a question and then the correct answer suddenly dawns on you, *do* change your response. But if you can choose only from a variety of guesses, go with your first or "gut" response.

Take Your Time

Yes, yes, yes, this is easier said than done. But the point is this:

- Plan your time.
- Work efficiently, but not in a panic.
- Make certain that your responses are legible.
- Take time to spell correctly. Even if an instructor does not consciously deduct points for misspelling, such basic errors will negatively influence the evaluation of the exam. Correct spelling of psychological terms is especially important —and can be especially difficult. Be careful.
- You may have to answer some questions involving statistics. Take time to check any mathematical work. Are the formulas correct? Did you calculate correctly?
- Take time to check your graphical work. Psychologists often use graphs, and exam questions may involve reading or drawing and labeling charts or graphs. Make sure what you draw is legible, and make doubly sure that you have labeled all features correctly and clearly. Some instructors who have contributed to this book report that they repeatedly observe careless errors of labeling. Points are lost for this.

- If a short answer is called for, make it short. Don't ramble.
- Use all of the time allowed. The instructor will not be impressed by a demonstration that you have finished early. If you have extra time, reread the exam. Look for careless errors. Do not, however, heap new guesses on top of old ones; where you have guessed, stick with your first guess.

Essay Exams: Read On

Exams in most introductory psychology courses are objective, consisting mainly of multiple-choice, fill-in-the-blank, and short-answer questions. A significant minority of survey-level courses, however, also include essay questions or paragraph-length short-answer questions on exams.

"*Essays!* That's for English and history. Psychology is about rats in mazes, not literature!"

The fact is that much of psychology concerns complex issues of thought, emotion, and behavior, and a growing number of instructors want you to *write* about psychological processes and issues. Therefore, please be sure to read Chapter 5 for advice on preparing for and writing effective essay responses.

CHAPTER 4

THINKING LIKE A PSYCHOLOGIST: KEYS TO WORKING THROUGH QUESTIONS IN PSYCHOLOGY

IT IS LITTLE WONDER THAT PSYCHOLOGY IS A POPULAR UNIVERSITY-LEVEL SUBJECT, EVEN among nonmajors. Who *doesn't* want to know more about human emotion and motivation? A knowledge of psychology, after all, should be useful in just about any aspect of life one can think of, including the arena of business as well as the most intimate aspects of one's personal life.

The good news is that just about all psychology survey courses do deal with issues of emotion and motivation, as well as with such other hot-button topics as mental illness and hypnosis, and even aspects of learning and memory that many students may find not only interesting, but quite helpful. However, if you enroll in an introductory psych course in the belief that the material is all about feelings, secrets of the mind, and generally what makes people tick, you're likely to be disappointed. Behavior, emotion, motivation, and so on are only three areas studied in the typical introductory survey. You will also likely look into areas that seem quite far afield from these topics, areas that include:

- Experimental methods
- Statistics
- Physiological psychology (an area that greatly overlaps biology)
- Sensation and perception
- Language

You will also find that much of an introductory course is not devoted to delving into the secrets of human behavior, but to defining and describing that behavior in scientifically useful terms.

This chapter promises to help you "think like a psychologist." But as should be apparent from what has just been said, psychology encompasses a number of quite diverse fields of interest. You will also learn that, even within a given area—human personality, say—a great many theories and approaches are current, many sharply differing from one another.

Does this mean, then, that it is impossible to think like a psychologist?

Not at all. As diverse as the field of psychology is, all psychologists do share certain basic assumptions and attitudes about their discipline. If you approach the course with a knowledge of these assumptions and attitudes, you can avoid surprise and disappointment regarding course content, and you can also increase your chances of performing successfully.

The Ground Rules

It is safe to begin with the assumption that the overall goal of psychology, however it is approached, is to understand the individual. It might be argued that this is the overall goal of most human beings. Certainly, most of us have opinions on what makes people feel, do, and behave in various ways. Many vocations and professions depend on some understanding of these things. We often speak of a talented teacher or a highly effective salesperson as being a "good psychologist."

And this is precisely the problem. The overall goal of psychology overlaps with concerns that most people share, regardless of vocation or level of education. Therefore, psychology introduces a second ground rule in addition to its goal of understanding the individual: conclusions about "the individual" must be based on scientific study—not just intuition or common sense.

Do you want to think like a psychologist? Then walk into the classroom prepared with the following mental equipment:

- **Open-mindedness:** Be willing to put your cherished ideas and perceptions on hold. Consider all of the new ideas that are presented to you. Give them a hearing. Think about them.
- **Skepticism:** Being open-minded does not mean blithely accepting every idea and observation that comes your way. *Question* and *challenge* what you hear and read. Evaluate it. Weigh it.
- **Objectivity:** The combination of open-mindedness and skepticism might be aptly described as objectivity. Recognize your biases (and those of others) as you evaluate the information you are given. Try to think about this information on its own merits, rather than based on your upbringing, cultural background, religious beliefs, and so on.

- **Systematic thinking:** While opinions on various aspects of emotion, motivation, behavior, and the rest are plentiful, few people attempt to assemble such ideas into a coherent system. However, this is precisely what psychologists try to do. You will benefit most from an introductory psychology course if you develop the habit of trying to understand not just isolated concepts, but concepts in systematic contexts. Freud's theory of the id is fascinating, but it is truly meaningful only in the context of how it functions in relation to the ego and the superego.
- **Provisional thinking:** If you go into the psychology course expecting to be given the answers to what makes people tick, you will be disappointed. Psychology is a collection of hypotheses and theories. A *hypothesis* is an informed guess or prediction about some aspect or aspects of whatever topic is at hand—human behavior under stress, for example. A *theory* is a systematic integration of hypotheses that have proven themselves reliable. Neither a hypothesis nor a theory is absolute fact or law. Indeed, it is the nature of hypothesis and theory that they cannot be proved definitively. We'll discuss this further in the next section, but the point here is that you must set aside any expectation of being handed the "right" answers, as if you were working a math problem. Rather, your job, as a psychology student, is to understand the many, often contradictory, hypotheses and theories that will be presented to you. Consider them all provisional and not final.
- **Thinking from multiple perspectives:** This is related to provisional thinking. In addition to being introduced to a variety of hypotheses and theories, you will be exposed to the various major schools of modern psychology. These typically represent different approaches to the same essential problems. Be aware that you need not embrace one of these schools and reject the others. Try to get inside all of them. Look at the world from the point of view of a psychoanalyst, a behaviorist, and a humanist.
- **Willingness to understand behavior from two perspectives:** Psychologists speak of two broad classes of behavior: overt behavior and covert behavior. *Overt behavior* can be observed and measured. Many psychological experiments deal exclusively with overt behavior. *Covert behavior* cannot be directly observed and measured. Mental processes, which are of great interest to many psychologists, are forms of covert behavior. It is important to understand and appreciate, from a point very early in the course, the distinction between these two forms of behavior.

Scientific Method

Historically, the fields of psychology and philosophy have been closely associated. By the later nineteenth century, however, psychology was clearly taking on a more scientific identity. In large part, this involved the psychologist's embrace of the scientific method. This approach is central to all scientific inquiry. As different as, say, physics is from psychology, both the physicist and the psychologist use the *scientific method* to guide their investigations.

The *scientific* method. This must be a most profound and complex method indeed!

Well, not really.

The scientific method is nothing more nor less than a systematic approach to a commonsense way of looking at the world—with the emphasis on *systematic.* It works like this:

1. A problem is stated.
2. Facts relating to the problem are gathered.
3. A solution to the problem is proposed: a hypothesis.
4. The hypothesis is tested.

Let's look more closely at these four steps. As the elements of the scientific method, they are the most important early steps you can take toward thinking like a psychologist.

The first step is to formulate a problem. For psychologists, the problem is typically a question about the effect of some factor or set of factors on some aspect of behavior. Like most scientists, psychologists are more interested in the mechanisms by which individuals operate than in questions of ultimate purpose. (Such questions are in the realm of philosophers and theologians, not scientists.)

Anyone—not just a scientist—can raise questions about the mechanisms of behavior, thought, and emotion. However, most of us don't, at least not consistently and systematically. Scientists may or may not be *born* curious, but it's a sure bet that they all *develop* curiosity. Without it, step one is impossible. You will increase your chances of excelling in psychology—and enjoying the course—if you work at cultivating your curiosity about the world and the people around you. Come to class and approach your textbooks with an attitude of curiosity, and learning will not only be easier, but much more meaningful.

After formulating a problem to study, the next step in the scientific method is to gather facts (data). Facts are gathered by

- Firsthand observation
- Measurement
- Counts

- Review of records of past observations

This material must be evaluated for reliability (Is it reliable? Why? Why not? *How* reliable is it?) and filtered for relevance to the problem. These processes typically require observational skill, an eye for detail, patience, a willingness to work hard, and objectivity. The latter quality is especially important. Data must be approached without prejudice. To prejudge the data is to compromise its usefulness.

Next, an educated guess is formulated concerning the relation between the problem identified and the facts gathered. This is the hypothesis. Its very special quality is that it *must* be falsifiable—that is, a statement capable of being proved false. It is *not* necessary that a hypothesis be proved true. Many, many common scientific hypotheses cannot be proved true beyond a doubt, yet they are still useful—as long as they cannot be proved beyond a doubt false.

The next step in the scientific method has as its object, therefore, not necessarily proving the hypothesis true, but proving it false. This is called *testing the hypothesis.* The scientist does his or her utmost to find ways to contradict or disprove the hypothesis. This is the phase of the scientific method that many people—not just in college, but throughout history—have the most trouble with. It is human nature to hang on to cherished beliefs, even in the face of evidence to the contrary. In the seventeenth century, when the astronomer Galileo introduced observational evidence supporting Copernicus's idea that the earth is not the center of the universe, he was put on trial by the Church, and came perilously close to losing his life. Similarly, ever since Charles Darwin developed the theory of evolution in the mid-nineteenth century, there has been no shortage of individuals and religious organizations who have opposed it.

Why is the theory of evolution, for example, called a *theory* rather than a hypothesis? A hypothesis that is repeatedly tested and never falsified is raised to the status of theory. Yet note that the theory of evolution has not been proved true. It has simply never been proved false. It has stood as a sufficient explanation of countless particular instances. It has never been successfully contradicted.

Testing of a hypothesis can be through experiment. An experiment is a test in which the researcher controls as many of the variables—the context and inputs of the test—as possible. Ideally, the effect of altering only one variable at a time is observed and noted. If feasible, a comparison is made between an *experimental group* and a *control group,* which are not different from one another except for the single variable being tested.

Where an experiment is not possible or practical, *observation* is used. Darwin, for example, did not try to experiment with evolution in a laboratory, but made meticulous observations of the fauna of South America and Africa. Sigmund Freud, the father of psychoanalysis, did not perform laboratory experiments on his patients, but observed them.

Watch Your Language

Just as you must get accustomed to the scientific method in your approach to psychology, keeping in mind its four basic steps, so you must understand that psychology has a vocabulary all its own. Essential to success in a psychology course is learning the vocabulary and then using it *accurately.* This requires memorization, but it also requires care and thought. Typically, psychological terms have limited and precise definitions. In contrast to everyday language, nuance and shades of meaning are avoided. Objective description and labeling are sought.

It is a good idea to create a running list of the technical and specialized words you encounter, complete with definitions *and* examples of *usage.* To be sure, your textbook will have a glossary of terms, and you will also find a basic glossary in this book. But such resources are no substitute for actually taking the time and effort to create your own list. Don't strive for completeness, but do make note of the terms you most frequently encounter, the terms that are obviously key.

While you should use the language of psychology with thought and care, you don't have to become paralyzed by fear of misusing terms. Just make certain that you understand each new term you encounter. Don't let any slip by. Such slips are cumulative. If you fail to understand term A, term B becomes that much more difficult to grasp. By the time you arrive at term C, you may be totally in the dark. Remember, too, that the first step toward study of any psychological problem or situation is *description,* and psychology provides an ample and precisely defined vocabulary to facilitate such description. The specialized terms of this science are the *tools* of the discipline. To work with psychology, even at the introductory level, you must have a sound knowledge of the basic tools. It's pretty hard to drive a nail with a screwdriver. Pick up the hammer instead. It's pretty hard to talk about the subject of *hallucination* if you think the word means the same thing as *delusion.*

Another List

In addition to your running list of specialized terms, you should compile a list of the major assumptions, theories, and psychological schools (as well as the major psychologists associated with them) that you frequently encounter or that are pointed out to you as especially important. Again, you will certainly find these in your textbook, but writing them down for yourself is a great, *active* way of learning them.

TIP: Think of this list as an annotated index. You should not only write down the particular law or principle, but make a note of where in your textbook it is discussed and explained.

Thinking Graphically

Many principles and experimental findings in psychology are not stated in words only, but also in formulas, tables, and even equations. Include these in your list. In some cases, graphic representation of the finding or principle is required. For example, sleep cycles are best understood in terms of a diagram. Don't overlook such illustrations. Work at interpreting and understanding them. You may even want to reproduce them in your list of principles.

In addition to diagrams of various sorts, you will also encounter a good many graphs in psychology. Resist the temptation to gloss over the graphs you encounter in your text. Instead *study* them—and by *study,* we mean *interpret:* think about the meaning of what you are seeing. The best way to go about this is to look at a graph and try to explain *in words* the relationships illustrated. This is not something you learn to do all at once. It takes practice—just as reading words on a page or notes on a musical score takes practice. There is nothing about C-A-T that *looks* like a cat, but, with practice, we learn to see a cat when we read *cat.* The same principle applies when working with graphs.

Math—in Psychology?

Some students choose psych to satisfy a core science requirement because they see it as a nonmathematical alternative to the likes of chemistry or physics. While it is true that those disciplines are sure to include lots of work with equations, psychology might also surprise you with some mathematical demands. Depending on the course, you may investigate such topics as cell function on the molecular level and neural transmission, which both involve some work with equations. You may also be asked to do some elementary work with statistics—a vital tool for interpreting the results of many psychological experiments.

Just as you must make an effort to understand the specialized terms of psychology, you should work on understanding any equations or statistical methods you encounter. Fortunately, these are all quite straightforward at the introductory level.

TIP: Most instructors suggest that you get into the habit of mentally translating equations into words. Don't just regard them as abstract sets of symbols. They describe important chemical and physical processes.

Pictures

In addition to graphs, diagrams, and equations, psychology includes a good many sketches and photographs of such things as nerve cells (neurons), the brain, sense organs, and experimental apparatus. These are not provided just to dress up the text. You should learn all that you can from the sketches and diagrams provided. Some exams may even call for you to sketch such basic structures as a neuron, so be certain that you are sufficiently familiar with the important images in the course.

Experimental Work

Many introductory psychology courses are taught exclusively in the classroom or lecture hall, but some include a lab section or at least some experimental or observational projects. If your course includes lab or experimentation, you can be certain that the work you do in these sections will show up on exams. Take notes, and be certain that you understand not just the process of the experiment or observational exercise, but the significance of it as well.

Be There

Psychology is a varied and complex subject. It is also an *eclectic* subject; that is, it borrows from various allied fields and disciplines, including biology, biochemistry, medicine, and statistics. Yet psychology is also an *orderly discipline.* It rests on clearly stated assumptions. It has clearly stated objectives and goals.

The key to studying successfully a subject that is complex, eclectic, and orderly is to proceed step by step, and to make certain that you understand each step before you move on to the next. Learn the vocabulary *well.* Understand the basic principles, theories, and laws *well.* Learn to work with graphs, diagrams, equations, and drawings *well.* Be certain that you understand all experimental or observational work fully, process as well as significance.

- Don't skip over textbook material.
- Don't just start reading words without taking the time and making the effort to understand them.
- Don't just glance at the graphs, diagrams, equations, and other illustrations you encounter. Interpret them by translating what you see into words.

Be there. When you study your textbook, *study it.* Practice the concepts you read about. Psychology is as much a set of observational skills as it is an intellectual pursuit. Acquiring and developing any skill requires active practice, not just passive reading.

Be there in class. Don't skip classes. Attend lectures. Take notes. Ask questions. Most instructors believe that *most learning takes place in the classroom.* This means actively listening to lectures, actively participating in class discussions, and—very important—asking questions *as they occur to you.* If you don't understand something in a lecture, ask about it as soon as possible. This is by far the most effective way of learning.

More Than a Set of Problems

We've all known people who are so intensely detail oriented that, in pursuit of every last fragment of minutiae, they consistently miss the big picture. As the old saying goes, they can't see the forest for the trees. Psychology, which requires learning many concepts and a great many new words, unfortunately presents this very trap. You can avoid falling into it by trying to think of psychology as something more than the sum of a series of concepts and words and diagrams. These are important, but they alone are not the purpose of psychology. You are getting a peek at a field that sets as its goal nothing less than understanding the individual in terms of emotion, cognition, motivation, perception, and behavior. Try to keep that in mind. It will drive and invigorate your work, and it will help you to bring together all those many constituent parts—vocabulary, tables, graphs, diagrams, drawings, equations —into a much more meaningful picture.

CHAPTER 5

THE ESSAY EXAM: WRITING MORE EFFECTIVE RESPONSES

MOST EXAMS IN INTRODUCTORY-LEVEL PSYCHOLOGY CONSIST EXCLUSIVELY OF multiple-choice and true-or-false questions, and many include such questions in addition to fill-in-the-blank segments and one-word or single-sentence short answers. A growing number of instructors, however, include essay questions in their exams. Often, these take the form of short essays, perhaps a paragraph or two in length. In some cases, however, they are more extended.

In this book, you'll find examples of all the psychology exam types you can expect to encounter.

Downside and Up

Griping is of little value at the outset of any enterprise, including a psych course that not only requires facility with memorizing and understanding a wide variety of concepts and perhaps even some skill with statistics, but also a way with words. Nevertheless, griping is human and natural, so we'd better get it out of the way.

Essay exams have the following distinct disadvantages:

1. They are intimidating. Even experienced, professional writers may get a shudder when they sit down to a blank page. Where do you begin? Worse, where do you go once you've begun?
2. Essay questions generally require deeper and broader knowledge of a subject than multiple-choice questions do.

3. Essay exams are time-consuming to take. It may be difficult to budget your time effectively.
4. Essay exams test not only your knowledge of course material, but your language and writing skills. This may seem like an unfair demand.
5. Essay questions may contain a significant element of subjectivity. Often, the issues are gray rather than black and white. Not only does subjectivity enter into your response, it also plays a role in the instructor's evaluation of the response. Instructors may respond to the skill of the presentation (or lack of such skill) as much they do to the substance of the answer. In some psychology courses, the instructor may actually *encourage* subjective essay responses.

In fairness, essay exams are almost as demanding on the instructor as they are on the students. They are much more difficult and time-consuming to grade than "objective" tests are. Instructors who use essay exams are demonstrating a genuine commitment to their students and their subject.

So there's the downside. But each of these negatives has a corresponding positive—if you know how to find and exploit it.

1. True, your blue book may be blank, but your mind doesn't have to be. First, there are effective ways to prepare for the questions on an essay exam—and we'll talk about these shortly. Second, take a good, long, careful look at the question. It should give you plenty to get you started. Get into the habit of using the terms, parts, and structure of the question as a kind of framework on which you construct the terms, parts, and structure of your answer.
2. If essay questions usually require deeper and broader knowledge of a subject than multiple-choice questions do, they also offer a deeper and broader stage on which you can play out your understanding of the course material. When you respond to short-answer and multiple-choice questions, you are limited by the instructor's rules: true, false, a, b, c, or d. It's a kind of binary situation. Either you *know* the *correct* answer or you don't—and if you don't, you lose. In responding to an essay question, you certainly need to address the question in all of its parts (don't stray, don't evade, don't get off the track), but you have much more control. You can focus on areas you know most about. You can play to your strengths and minimize your weaknesses.
3. Essay exams are time-consuming to take. That's a fact. And instructors know it. They take into account the pressure of limited time and the fact that you are writing a single draft when they evaluate the essay. This generally prompts them to overlook a lot of sins of omission and even outright errors. Time pressure actually *reduces* instructors' expectations.
4. If essay exams test not only your knowledge of course material, but also your

language and writing skills, it behooves you to polish those skills. Good writing will earn extra points. It's that simple. You may not know more about psychology than the person sitting next to you, but if you write more effectively, you will earn a higher grade. An added bonus: The more clearly and effectively you can express yourself, the better your own understanding of the material you are writing about will be. Effective writing not only communicates knowledge to others, it helps you to communicate with yourself.

5. Essay questions contain a significant element of subjectivity, it is true, and this can give you that much more room to be right. Create a skillful presentation, and you are likely to be evaluated positively, even if you miss some issues.

Study and Preparation

Even if they deny it, most instructors tend to "teach for the midterm and final." More accurately, they construct exams that genuinely reflect the course content, including particular themes and topics that are emphasized. Do instructors ask trick questions? Rarely. Do they deliberately try to mislead you, hiding exam material in the background of the course, as if it were an Easter egg? Almost never. Instructors want you to succeed. Good test performance tells them that they have gotten through to you. The more students who do well, the more successful an instructor feels. This being the case, be certain to take careful lecture notes. Make a good set of general notes, but also listen selectively for:

- Points that stand out
- Points that are repeated
- Points preceded by such statements as, "Now this is important" or "This is a major issue."

Assume that any point, theme, or topic that is given special emphasis will appear as an exam topic. The more emphasis it is given, the more likely it is to appear as an essay exam question.

Typically, course lectures mesh with textbook assignments, additional assigned outside reading, and perhaps class handouts. Take notes on all of these sources. Handouts are usually especially important. Assume that handout material will figure in some way on any exam.

Some introductory psych courses include lab or experimental work. Be certain that you know just what part any lab exercises or experimental projects will play in the exam. Essay questions may well focus on lab work or an experimental project, asking you, for example, to describe the experiment, to describe its results, and to draw conclusions about its significance. Indeed, lab work and special projects are

TIP: Some instructors prepare examinations well ahead of time, but most write them up shortly before they are given. Usually, instructors will review their notes in preparation for writing exams. This makes it even more likely that well-emphasized subjects and issues will appear on the exam.

natural candidates for a narrative essay question. Be certain that you take good lab or experiment notes and that you do not neglect to study them.

Avoid passivity. *Ask* the instructor to talk about the scope of the exam. Also seek out students who have taken the course before. Ask them about the exam. Many instructors have favorite themes or concerns, which get repeated from year to year.

TIP: Do not confine your preparation to your notes. If past course exams are available for your review, review them. Use them as practice tests. Of course, you should also make use of the exams in this book. These days, many instructors post past exams on a special Web site devoted to the course.

You Are Not Alone

You needn't face the exam alone. Group study is often a highly effective way of preparing for essay exams. Indeed, many instructors encourage students to form study groups. This is because group discussion tends to bring out major themes and issues, the meat and potatoes of essay exams.

Don't make the mistake of allowing the study group to dissolve into a social hour. Consider focusing the discussion by having each member of the group make up a sample essay question. Use these as the topic of discussion.

PITFALL: By all means, examine tests from previous semesters, but don't make the mistake of assuming that these will simply be repeated this semester. Most instructors change tests from year to year. Examine past tests to get an idea of the type and scope of questions asked—not to get specific answers.

Limited Possibilities

Let's assume you're not musically inclined. Now, look at the score of, say, Beethoven's *Moonlight Sonata.* All those notes! All those chords! How does anyone ever learn how to read so much simultaneous information, much less translate it into sound on a keyboard?

Well, it requires study, hard work, practice—and inborn musical talent helps, too. But it really isn't as hard as it appears to the unmusical. While, theoretically, there is an infinite number of ways in which musical notes can be combined and deployed, in actual practice, the possibilities are indeed limited. Most chords and note sequences occur in recognizable groups and patterns. Just as, when you read a book, you don't spell out individual words or struggle to recognize individual letters, but instead more or less unconsciously interpret familiar linguistic patterns and phrases, so musicians process the notes they see.

Now, you may think that the range of questions possible in an essay exam is virtually infinite, but, actually, like the notes of a musical score or the words on a novel's page, the range is limited—and this is true regardless of subject.

There are a limited number of questions that can be asked about any theme, subject, or topic. Furthermore, each question is controlled by a key word. Knowing those key words—and understanding their meaning—will help you to prepare adequately for the exam. Here they are:

Analyze	Literally, take apart. Break down a subject into its component parts and discuss how they relate to one another.
Compare	Identify similarities and differences between (or among) two (or more) things. End by drawing some conclusion from these similarities and differences.
Contrast	Set two (or more) things in opposition in order to bring out the differences between (or among) them. Again, draw some conclusion from these differences.
Criticize	Make a judgment on the merits of a position, theory, opinion, or interpretation concerning some subject. Support your judgment with a discussion of relevant evidence.
Defend	Give one side of an argument and offer reasons for your opinion.
Define	State as precisely as possible the meaning of some word, phrase, or concept. Develop the definition in detail.
Describe	Give a detailed account of something. Where biology subjects are concerned, this account will typically be step-by-step, with the emphasis on cause and effect.
Evaluate	Appraise something, rendering a judgment on its truth, usefulness, worth, or validity. Support your evaluation with relevant factual evidence.
Explain	Clarify something and provide reasons for it.
Identify	Define or characterize names, terms, things, places, events, or other phenomena.
Illustrate	Provide an effective example of some stated point, principle, or concept.
Outline	Show the main features of some event, concept, idea, theory, or the like. Omit the details. Often, such an answer will be in outline rather than narrative form.
Pros and cons	A more specialized form of *evaluate.* List and discuss the positives and negatives about a certain position, idea, event, theory, or the like.
Relate	Narrate an observation or set of observations; emphasize the relation of one observation to another, especially focusing on cause and effect.
Review	Survey a subject. This is much like an *outline*, but put in narrative rather than list or outline form.
State	Present your answer briefly and clearly, usually using a simple declarative sentence: "Such-and-such is such-and-such."

Summarize Give a concise account of major points, ideas, or events. Skip details and examples.

Trace Follow an event, theory, or idea to its origin. The form of this response is generally: "The origin of A is X, Y, Z." Then the rest of the answer continues by elaborating on X, Y, and Z and moves forward to A.

One or more of these are the intellectual operations basic to just about any essay. Look for these key words in the essay question.

Many of these operations focus on just two elements:

- Cause
- Effect

Indeed, most psychology essay questions deal with causes and effects. Be prepared in advance to work with both elements.

TIP: You can use the operations and elements just discussed to focus your study notes. For example, instead of listing a bunch of unrelated facts about the subject of neural transmission, why not focus your notes in terms of causes and effects and also *trace* particular effects to their causes? You might study neural transmission by *outlining* all of the elements operative in it, including the general structure of a neuron, the electric potential of a neuron at rest, how that potential changes when the neuron is stimulated, the function of axon and dendrite, and what happens at the synapse.

Test Time: The Problem of Inertia

If you've just finished an exam in Physics 101, maybe you recall what Sir Isaac Newton had to say about inertia—the tendency of a body in motion to remain in motion and a body at rest to remain at rest. Sitting down to an essay question on exam day, many students are confronted by their own personal form of inertia. You look at the question. And look at it. And look at it some more—hoping, perhaps, that the letters will magically rearrange themselves on the page to yield the answer.

The bad news, of course, is that they will do no such thing. But the good news is that the germ of the answer is, in fact, in the question. How do you overcome essay test inertia—that paralyzing difficulty in getting started? Just read the question. *Really* read the question.

In fact, don't worry about writing just now. Sit down and read *all* of the questions before you begin to write anything.

Here's why:

- If you are given a choice of which questions to answer (say two out of three), you want to be sure that you answer the ones you know best.
- Questions are sometimes related. You want to get an idea of just how they are related to one another, so that you don't waste too much of your answer on one question to the neglect of another.
- You need to assess your time needs. Are there some points that will require more time than others?

- You want to make certain that you answer the questions you are confident about first. Given a limited amount of time, be certain that you get to your best shots first and complete them before attending to the questions you're less confident about.

Before you begin to write, be absolutely certain that you understand each question completely. Read each question actively, aggressively, with pencil in hand:

- Identify and underline key words—including those listed above as basic operations and elements.
- Be certain to *do* what the key words ask. For example, don't just *define* when you are asked to *explain.*
- If a question is complex, consisting of several subquestions, make certain you understand and answer all parts of the question.

Generally, you should let the question provide the basic structure for your response. If a question consists of subquestions A, B, C, and D, begin by answering A, then B, C, and D. If you have a *very good* reason for changing the order of your response, be certain to explain it. For example, you might begin: "Because C is essential to understanding A, B, and D, I will begin by discussing C, then proceed to A, B, and D."

Finally, answer the question—and only the question. Make certain that you address all parts of the question, but don't go beyond what the question asks, unless, after you have thoroughly addressed each aspect of the question, you feel it is important to bring in additional issues. If you do so, tell your reader what you are doing, so that he or she won't think you've misunderstood the question and are simply going off on a tangent.

TIP: People who give a lot of speeches are fond of offering this formula for a successful speech: Tell them what you are going to say. Say it. Tell them what you said. You might keep this in mind when you are answering an essay question, though you should elaborate on the formula a bit:

1. **State your subject or thesis.**
2. **Briefly state how you will discuss it. (A, B, C, D—or maybe C, A, B, D.)**
3. **Answer the question.**
4. **Concisely summarize your answer.**
5. **Draw any additional conclusions as to significance, ramifications, and so on.**

To Outline or Not to Outline?

Sometimes, pulling out ideas in response to an essay question is difficult, halting, and laborious. Sometimes, however, you are flooded with ideas. Either situation can make inertia more powerful and, at worst, bring on mental paralysis or, at the very least, cause you to write a poorly organized essay. To prevent these outcomes, apportion your time so that you spend about as much time planning and outlining your response as you do actually writing it.

Now, an essay exam outline does not have to be a formal outline with major and minor headings. Perhaps a simple list will be sufficient. Just make a map of your answer, setting down the main points that need to be covered. This will have three effects:

1. It will help ensure that you leave out nothing important.
2. It will help you organize the logic of your response.
3. It will reduce your anxiety.

The first two points are obvious. The last is less obvious, but no less important. Without an outline, you may fear that you will forget something important or get lost in your response. Get the main points out of your head and onto paper quickly, and you won't have to worry about forgetting anything or getting lost.

Structure Strategy

Make the structure of your essay exam response as clear and obvious as possible. Begin with a thesis statement: a statement or listing of the main ideas you will support and develop in the body of the essay.

Where do you get your thesis statement?

The first place to look is the question: "Explain what the diathesis-stress model of mental illness is (i.e., vulnerability-stress) and give an example of research data that supports this model."

There's the question. Begin by getting to the point: "The diathesis-stress model of mental illness is divided into two parts: a biological factor and a trigger factor. The point of the model is to show that, due to a trigger in predisposing (biological) factors, depression can remain constant or variable."

Note how the first sentence in the response briefly *explains,* which is just what the question called for. The second sentence contains the thesis by proposing what the *point* of the model is. This sets up the content of the rest of the essay, which, as asked for in the question, consists of an *example* that supports the model (which is also the thesis of the student's essay).

After supporting the thesis with an example (see Answer B to question 16 in Chapter 21), the student draws a conclusion: "Therefore predisposing factors may inhibit long-term depression in people." We know this is a conclusion because the sentence begins with the "signpost" word *therefore.*

In general, make the thesis statement as simple and as direct as possible. State the thesis. Present your plan for supporting your thesis. In this case, your plan is also a simple one: you will provide what the question asks for, supporting the thesis with an example.

Follow your plan in the body of the essay:

I. Thesis: "The point of the model is to show that, due to a trigger in predisposing (biological) factors, depression can remain constant or variable."

II. Plan: "Here is an example . . ."

III. Present the example.

A. Narrate the example.

B. Highlight a way in which the example supports the thesis.

C. Highlight another way.

D. And, if possible, highlight yet another.

IV. Conclusion: Make the "therefore" statement.

The KISS Formula

Can you really get ahead by giving your professor a *KISS*? Well, sort of. This acronym stands for *K*eep *I*t *S*imple, *S*tupid.

Now, let's get something straight. This does not mean that you should overly simplify complex ideas or issues, let alone avoid them. But it does mean that you should structure your presentation of ideas in as simple a form *as possible.* This means:

- Be concise.
- Be direct.
- Start the essay with a thesis statement.
- Start each paragraph with a topic sentence, announcing the subject of the paragraph.
- Try to make a single major point in each paragraph.
- When you move on to a new point or new idea, start a new paragraph. Essay exam essays usually have short paragraphs—certainly shorter than what you'd write in a term paper, for example.
- Draw a definite conclusion.

Specify, Always Specify

Wherever possible, avoid abstraction. In place of vague generalizations, make very specific points that use specific examples. Examples are important in any essay response, and they are especially important in essays on psychological subjects. Remember, science begins with observation—that is, science *begins* with examples.

Signposts

Develop a repertoire of verbal signposts. One of the most effective signposts is enumeration. For example, instead of saying "The scientific method consists of *several* steps," write "The scientific method consists of *four* major steps." Then go on to list and discuss all four steps.

Enumeration accomplishes three things:

1. It creates the impression that you are in control of the information.
2. It creates an impression of precision and completeness.
3. It sets up an expectation in the reader, who is satisfied when that expectation is fulfilled. You promise four items. You deliver four items. The reader is impressed.

Other signposts include words and phrases such as:

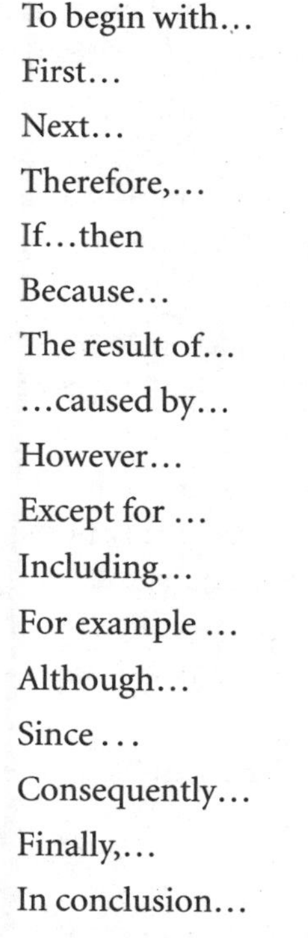

To begin with...

First...

Next...

Therefore,...

If...then

Because...

The result of...

...caused by...

However...

Except for ...

Including...

For example ...

Although...

Since . . .

Consequently...

Finally,...

In conclusion...

Use these to get your reader from one point to the next, to make clear exceptions, and to point your reader toward your conclusions.

A Few Words on Words

Take time to choose your words carefully. This does not mean trying to impress the instructor with big words or fancy words, but do try to find the *right* words, the words that most precisely say what you mean.

TIP: The language of psychology is typically far more precise than everyday language concerning emotion and behavior. Make certain that you use it correctly. For example, in everyday speech, we might use the words *attitude* and *belief* interchangeably. To a psychologist, however, *attitude* is a relatively enduring idea about people or things that has a significant emotional component, whereas a *belief* is such an idea without the emotional component. The difference between the terms is significant.

- Use the language of psychology. Identify and become comfortable with the specialized terms used in lectures and textbooks. Understand them thoroughly. Use them appropriately in writing the exam essays.
- Avoid slang. Slang is not only imprecise, it creates a poor impression.
- Prefer strong, precise nouns and verbs to adjectives and adverbs. This will help you to convey more accurate meaning. From the point of view of a psychologist, the sentence "IQ is a measure of intelligence" is a significantly weaker statement than "As originally used, IQ was a number based on the ratio of mental age to chronological age. It was conceived as a measure of intelligence."
- Express yourself in a direct manner. Don't load up sentences with unnecessary words. Try to make each word count. Use the active voice instead of the passive voice: "In psychoanalysis, catharsis is a venting of emotion that relieves underlying pressures" is a much stronger sentence than "A venting of certain emotions is what psychoanalysts often call catharsis."
- Avoid unnecessary qualifying phrases and waffling words such as "it has been said" or "I think" or "it seems to me" or "it seems likely that." Make direct statements.
- Avoid padding and repetition like this: "Introspection is all about self-examination, looking into yourself, really thinking about yourself, and then coming up with an analytical description of what you are really, really thinking or feeling." Instead: "Introspection is looking within and describing what you think or feel."

TIP: Try to budget time to reread and proofread your essay. Catch and correct errors of usage, grammar, and spelling.

- Avoid errors of usage, grammar, and spelling. If you have trouble in these areas, work on them. Such errors undercut your credibility.

Neatness Counts

Take enough time to write legibly. Make your work as easy to read as possible. If the instructor has to struggle to decipher your handwriting, he or she will easily lose the thread of your discussion, and your grade is likely to suffer as a result.

Recycle!

When the graded exam comes back to you, resist the temptation either to pat yourself on the back or to kick yourself in the backside. Instead, carefully read the examiner's comments. Learn from them. Schedule a conference to discuss the exam—*not* with the goal of getting your grade changed, but of identifying those areas that can use improvement.

TIP: Some psychology essay questions call for you to illustrate your answer, typically with a graph or simple flowchart. You don't have to produce great works of art, but be sure that your illustrations are legible, with all features clearly—and correctly—labeled. Slow down. Take enough time to be neat.

When you have a conference, try not to respond defensively. Invite frank feedback. Don't get offended or upset. Instead, look for patterns. Does the instructor say that you simply failed to answer the question adequately? That you didn't answer all parts of the question? That you answered vaguely? That you punctuated poorly? That you didn't use enough examples? Diagnose areas that need improvement, even as you identify your strengths. Your ultimate goal is to avoid repeating errors while working to duplicate your successes.

PART TWO
STUDY GUIDE

CHAPTER 6

INTRODUCTION TO PSYCHOLOGY: THE MAJOR TOPICS

WHILE THE WORD *PSYCHOLOGY* MEANS, IN GREEK, THE "STUDY OF THE MIND," IT IS more accurate to describe modern psychology as the study of the mind and the relationship of mind and body. It is also true that while psychology is primarily interested in understanding the individual, its research methods very often involve the study of groups. Because of the mind-body focus and the work with groups, modern psychology straddles—and draws heavily on (as well as contributes to)—biology and the social sciences. Indeed, in some universities and colleges, psychology is classified as a natural science, while in others it is viewed as a social science.

Psychologists study diverse issues such as learning, cognition, intelligence, motivation, emotion, perception, personality, mental disorders, and behavior genetics (the study of the age-old "nature-nurture" question: the roles played by genetic inheritance versus environmental factors in shaping the individual).

Broad Specialties

Clearly, the reach of psychology recognizes few boundaries, with the result that the field often seems confusing and poorly defined. To be sure, psychology is not unified by a single theoretical structure, and the fact that it is freely divided between applied fields (such as clinical psychology) and experimental fields adds to the apparent welter that characterizes the discipline. Moreover, while the primary and ultimate focus of psychology is on human beings, many experimental psychologists work exclusively with animals ranging from single-celled organisms to primates. Applied psychology fields include:

- Clinical psychology
- Counseling psychology
- Industrial psychology
- Consumer psychology
- Environmental psychology

Of these, the most important is clinical psychology, which deals with the diagnosis and treatment of mental disorders. Businesses use industrial psychology to aid in employee selection and related issues, and they call on the field of consumer psychology to enhance marketing programs and improve products. Environmental psychologists study how living and work spaces affect the general quality of life, productivity, and well-being.

Experimental psychology includes:

- Child psychology (a field devoted to the study of children and not to be confused with a subspecialty of clinical psychology devoted to mental disorders of children)
- Educational psychology
- Social psychology
- Physiological psychology
- Comparative psychology

Child psychology applies psychological theory and research methods to children, while a related field, educational psychology, studies learning processes and learning problems. Although much psychological work is based on studies of *groups*, such studies are usually used to generalize about the *individual*. Social psychology, however, studies groups and is concerned with group dynamics as well as human behavior in a social and cultural context. In this, it is often closely allied with sociology and anthropology. Comparative psychology compares behavior across animal species and is closely allied with biology and zoology.

The Schools

Psychology is not only divided into formally specialized fields, it is also dominated by what have been variously called "schools" or "forces."

Psychoanalysis

In modern psychology, the first great school was psychoanalysis. So influential has this school been that many people think of *psychoanalysis* and *psychotherapy* as one and the same. In fact, psychoanalysis is a specific approach to human behavior and to the diagnosis and treatment (i.e., psychotherapy) of mental disorders.

The father of psychoanalysis was Sigmund Freud (1856-1939), one of the most influential thinkers of modern times. A brilliant Viennese physician, Freud based psychoanalytic theory on his clinical work. He concluded that the human mind is tripartite in structure, consisting of the id (seat of instinct and the ultimate driver of all motivation), the ego (essentially equated with the conscious self), and the superego (essentially the moral self). The extraordinary implications of this tripartite structure are discussed in Chapter 15, but suffice it to say here that Freud introduced the idea of the unconscious mind as a primary determinant of personality and behavior. Psychoanalysis is the most elaborately systematized of current personality theories, and the psychoanalytic school is perhaps the single most influential force in modern psychology.

Some of the most important modern psychologists have been inspired by psychoanalytic theory, though many have sought to take it in new directions. Some of the following will be discussed further in later chapters:

- Carl Jung (1875–1961) was a disciple of Freud, who later expanded his mentor's theory of the personal unconscious into a theory of the *collective unconscious*, an entity shared by all human beings and containing *archetypes*, which are universal forms, images, and ideas that motivate religion and mythology.
- Alfred Adler (1870–1937) was another of Freud's students, who went on to define a "striving for superiority" as the prime mover behind the development of personality.
- Erik Erikson (1902–1994), yet another Freudian, expanded his teacher's theory into the realm of social interaction.
- Karen Horney (1885–1952) started from Freud's major assumptions, but placed greater emphasis on conscious processes in shaping personality. She also revised Freud's explanation of differences between the male and female personality.

Understanding psychoanalytic theory can be challenging. Often, it resembles philosophical and even theological discourse as much as it does science. However, the effort is well worth it, since so much of contemporary literature, art, and culture—in addition to psychological thought—is influenced by or even based wholly upon psychoanalysis.

Behaviorism

Whereas Freud and his followers theorized about the inner processes of mind, certain psychologists rejected analyzing what cannot be directly observed. Instead, they insisted on studying only *overt behavior*, which can be directly observed. For the strict behaviorist, the only empirical—measurable—data consists of a *stimulus* and a *response*, input and output. A stimulus is any sensory event that can be detected, and a response is any behavior the subject is capable of performing. The *mind* was regarded as a kind of black box, the workings of which were essentially unknowable.

While this may seem a limited and limiting view, the behaviorists argued that, actually, stimulus and response are all that is necessary to understand and explain behavior.

Behaviorism is surveyed in more detail in Chapter 12; important names in the field include:

- E. L. Thorndike (1874–1949), who formulated the cornerstone Law of Effect, which holds that animals repeat behavior that results in reinforcement and do not repeat behavior that results in punishment.
- Ivan Pavlov (1849–1936), a Russian physiologist who studied *conditioning*, using dogs. Pavlov explored in detail the relationships between stimuli and response.
- J. B. Watson (1878–1958), the principal proponent of *stimulus-response* (S-R) *behaviorism*. Watson believed that environment far outweighs genetic inheritance in shaping human personality.
- B. F. Skinner (1904–1990), who greatly extended S-R behaviorism through the concept of *operant conditioning*, which introduced a volitional, voluntary element into the conditioning process. Skinner believed that the model of operant conditioning could be used ultimately to explain all behavior, including advanced, complex human behavior.

Humanism

As some psychologists saw it, both psychoanalysis and behaviorism were insufficiently satisfying in human terms to account for personality and distinctly human motives. The humanists argued that both of these schools emphasized the animal aspects of human behavior at the expense of the human aspects.

Humanists such as Carl Rogers (1902–1987) and Abraham Maslow (1908–1970) argued that human beings are motivated by uniquely human issues, including needs to share, to belong, to help others, and to *self-actualize*—to realize one's own potential.

Gestalt Psychology

The German word *Gestalt* means the way a thing has been "put together," and

Gestalt psychology, which was first developed by the German psychologist Wolfgang Köhler (1887–1967), holds that the sum of anything is greater than its parts; that is, the attributes of the whole cannot be inferred from analysis of the parts out of context. Like humanism, the Gestalt view sought to humanize psychology by refuting attempts to reduce the understanding of behavior to a binary system of stimulus and response. Gestalt psychology became very influential and forms the basis for many modern theories of perception (Chapter 13).

Cognitive Psychology

Yet another, more recent, rejection of behaviorism is the cognitive view, which deals with problem solving and thinking as opposed to mere patterns of stimulus and response. A number of contemporary psychologists use computer science and information-processing theory to model and explain cognition. (See Chapter 13.)

Physiological Psychology

The field of physiological psychology, closely allied to the biological and medical sciences, has made many recent advances in exploring and explaining physiological processes in the brain, actually mapping, for example, the areas of the brain in which various functions of perception and emotion are concentrated.

Road Map

Remember, this book is a guide to help you use midterms and finals (inevitable facts of college life), to focus study of your psychology course. It is not a comprehensive introduction to psychology, and it is certainly not a substitute for reading your textbooks, attending lectures, and actively participating in class discussion. This chapter and the others in this part of the book should serve to point out the most prominent psychology landmarks you will encounter on your trip through the course. We have already inventoried some of the most basic assumptions and principles of psychology. Here are the major themes and topics that are built on these assumptions and principles and that are reflected in the next eleven chapters.

History of Psychology

- The Greeks
- The Middle Ages
- The Seventeenth and Eighteenth Centuries
- The Nineteenth Century
- Functionalism
- Structuralism

- Psychoanalysis
- The Twentieth Century
- Behaviorism
- Gestalt Psychology
- Humanistic Psychology
- Cognitive Psychology
- The Eclectic Approach

Research Methods

- Experimental Methods
- Operational Definitions
- Equivalence of Procedures and Groups
- Blinds
- Statistical Analysis
- Nonexperimental Research
- More About Statistics: Correlation

Physiological Psychology

- Nerve Cells and Nervous System
- CNS and PNS
- The Senses
- The Brain
- Mapping the Brain
- Forebrain Structures
- Cerebral Structures
- Sensation and Perception
- Measuring Sensation
- Visual Sensation
- Auditory Perception
- Taste and Smell
- Touch and Other Senses
- Endocrine System
- Pituitary Gland

- Thyroid Gland
- Parathyroid Glands
- Adrenal Glands
- Pancreas
- Ovaries and Testes
- Pineal Gland
- Thymus Gland
- Other Endocrine Secretions
- Genetics
- Predicting Genetic Possibilities
- Crossovers and Mutations
- Sex-Linked Traits
- Action of Multiple Alleles
- Importance of Genetics to Psychology

Developmental Psychology

- Stages and Periods
- Prenatal Development and Psychology
- Infant Development
- Two Key Developmental Theories
- Piaget's Cognitive Theory
- Development of Moral Reasoning
- Critical and Sensitive Periods
- Erikson's Life-Span Theory

Consciousness, Emotion, and Motivation

- The Levels of Consciousness
- Waking
- Subconsciousness
- Selective Attention
- The Unconscious
- Sleep States
- Rhythms and Stages

- Dreams
- Sleep Disorders
- Hypnotic States
- Drugs and Consciousness
- Emotion
- Motivation
- Primary Motivation Disorders
- Sexual Motivation

Conditioning and Learning

- Conditioning
- Operant Conditioning
- Applied Conditioning
- Learning Theories
- Synthesis

Cognitive Processes

- Memory
- More About Memory Types
- Forgetting
- Thought
- Language
- Acquisition and Development of Language

Intelligence

- Definitions
- IQ
- Other Test Issues
- Special Cases
- Retardation
- Learning Disorders
- Giftedness
- Can IQ Be Enhanced?
- Genetics Versus Environment

Personality

- Psychoanalysis
- The Tripartite Structure of Personality
- The Psychosexual Stages
- The Neo-Freudians
- Jung
- Adler
- Humanistic Theories of Personality
- Personality as Traits

Mental Illness

- What's Abnormal?
- What Is Unusual or Bizarre?
- What Is Maladaptive?
- What Is Troublesome?
- Types of Disorders
- Major Categories: A Closer Look
- Mood Disorders
- Anxiety Disorders
- Dissociative Disorders
- Schizophrenia
- Personality Disorders
- Treatment of Mental Illness
- Historical Perspectives
- Approaches to Treatment

The Psychology of Groups

- Principles
- Attitude Change
- Persuasive Appeals
- Indirect Causes of Change
- Conformity
- Compliance
- Obedience
- Observational Learning

CHAPTER 7

HISTORY OF PSYCHOLOGY

MODERN PSYCHOLOGY MAY BE SAID TO HAVE BEGUN WITH THE WORK OF SIGMUND Freud, whose practice of *psychoanalysis* was the first attempt to understand mind and the relation of mind and body from a truly comprehensive and scientific point of view. While Freud's intellectual background was highly literary and philosophical, he was a physician treating patients, and his theories were all based on and supported by his empirical experience with those patients as well as on his own self-analysis.

But if fully systematic scientific psychology began with Freud, he was by no means the first student of psychology. Indeed, psychology is one of the oldest of humankind's interests, and it is likely that, at some point early in your introductory course, you will review the history of psychology before Freud.

The Greeks

It is no accident that the word *psychology* combines the Greek words *psyche* (soul or mind) and *logia* (study). Both Plato and, even more widely, Aristotle speculated widely on the nature of mind. Aristotle believed that the mind or soul was separate from the body, and this *dualistic* view came to dominate Western ideas about how the mind and body are related. From Aristotle has come the notion that the mind—or soul—is a kind of spiritual entity, divinely endowed with the capacity for reason and virtue. Mind, stands in stark contrast to the material body, imperfect and corruptible—in every sense of the word.

Aristotle did more than speculate on the role of mind and its relation to the body. He thought intensively about how the mind functioned, concluding (as had Plato before him) that human beings perceive, learn about, and come to understand by forming mental associations between the events we observe. This principle—later called *associationism*—stated that recalling a past event or experience brings into use other events and experiences that have become related to this event in one or more ways. Aristotle spoke of three major kinds of association—similarity, contrast, and contiguity (two events are associated if they occur more or less at the same time)—and, over time, his speculations were elaborated by others into an understanding of just about all aspects of mental life, save original sensations. Indeed, many aspects of associationism, particularly contiguity, continue to figure importantly in theories of learning and memory.

The Middle Ages

While the European Middle Ages turned its back on much classical learning, it embraced the philosophy of Aristotle, especially his dualistic view of mind (or soul) and body, which harmonized well with Christian doctrine. While much that we would call psychology merged with theology during the Middle Ages, the period also saw the emergence of what might now be called *humoral psychology* or *humoral physiology.* The belief that the proportions of the four *cardinal humors*—blood, phlegm, choler (yellow bile), and melancholy (black bile)—in the human body determined personality and behavior is of ancient origin but was especially pervasive during the Middle Ages and even later. The ideal person enjoyed perfect balance among the four humors; however, few people were so fortunate. If blood predominated, a person would be sanguine—overly passionate. If phlegm predominated, he or she would be phlegmatic—slow and lazy. A preponderance of choler created an angry personality, while too much melancholy brought on depression.

The Seventeenth and Eighteenth Centuries

During the renaissance, a number of thinkers and *natural philosophers* took a fresh look at psychology. The French philosopher and mathematician René Descartes (1596–1650) devoted a great deal of attention to the question of the mind's relationship to the body, concluding that the two are indeed separate, but nevertheless interact, thereby determining who we are, what we know, and how we behave. Moreover, Descartes espoused a view that would come to be known as *nativism;* he believed not only that people were born with the ability to think and reason, but that heredity was primarily responsible for who we are, for personality, for talents, even for moral inclination, and that learned or acquired characteristics played a comparatively minor role.

During the seventeenth and eighteenth centuries, the Englishmen Thomas Hobbes (1588–1679) and John Locke (1632–1704), the Scots thinker David Hume (1711–1776), and the Irishman George Berkeley (1685–1753) reacted against the Cartesian nativist view. These philosophers, called empiricists, argued that the mind was essentially empty at birth. Locke, the most extreme of the *empiricists,* called the infant mind a *tabula rasa,* a blank slate. As Locke saw it, nothing in intelligence or personality was innate, and all knowledge of the outside world was acquired through the senses, through experience, which leaves its mark on the "blank slate."

The implications of the nativist versus the empiricist view are profound. The nativist view leaves little room for free will or even moral responsibility. We are what we are, and there is nothing that can be done to change in a significant way what we are born with. The empiricist view, in contrast, puts a great burden of responsibility on the individual and even more of a burden on society, arguing that we are born amoral and that personality, as well as moral character, is molded by *nurture.* It is the responsibility of the family and of society at large to provide an environment that will create desirable traits in individuals.

The debate between the roles of "nature" and "nurture" in the formation of personality and character continues today. At the end of the eighteenth century, the group of writers and philosophers who created the Romantic movement in the arts and philosophy argued an extreme nativist position in reaction to the empiricism of Locke and the others. The French philosopher Jean Jacques Rousseau (1712–1778) presented the most extreme view, arguing that people were born inherently good, but that society corrupts us.

The Nineteenth Century

Through the eighteenth century, psychology partook more of philosophy than of science. In the mid-1800s, however, two German scientists, the physiologist Johannes P. Müller (1801–1858) and the physicist-physiologist Hermann L. F. von Helmholtz (1821–1894), began to bring the field into the fold of the sciences. They conducted experimental work on sensation and perception, Helmholtz going so far as to measure the velocity of impulses through nerves. (Using a myograph, which he had invented, Helmholtz clocked impulse velocity at 90 feet per second.) Not only did the work of Müller, Helmholtz, and their various students lay the groundwork for modern physiological psychology, it demonstrated the practical possibility of scientifically studying the physical processes behind mental activity.

Functionalism

Another spur to the transformation of psychology from philosophical speculation to science came with the work of Charles Darwin (1809–1882), the creator of evolutionary theory and the theory of natural selection. Darwin's work showed that organisms evolve over many generations by adapting to the demands of the environment. Adaptations that promote survival tend to be passed on, while those that fail to promote survival ultimately disappear as fewer organisms with such non-adaptive characteristics survive to reproduce.

Inspired in part by Darwin's findings, the American psychologist William James (1842–1910) developed the theory of *functionalism,* founded on the idea that our personalities, character, and aptitudes are strongly shaped by adaptation to the environment. This view was significantly different from Darwin's natural selection theory in that James focused not on the evolution of a species over time, but on the "evolution" of an *individual* personality in the course of *individual* development. Still, *functionalism* required psychology to study both the environment and the individual, a cornerstone idea on which much of modern psychology still rests.

Structuralism

In 1875, at Harvard University, James founded perhaps the world's first psychology laboratory. In 1879, Wilhelm Wundt (1832–1920) founded a similar laboratory at the University of Zurich. Wundt sought to investigate the immediate experiences of consciousness, including sensations, feelings, volitions, and ideas. His chief experimental method was *introspection,* the conscious examination of conscious experience. Introspection is still used in experimental psychology and, indeed, is probably the most important treatment method in clinical psychology.

Wundt's approach was based on the assumption that human conscious experience could best be understood by analyzing it—that is, breaking it down into its component parts. One of Wundt's students, the English-born psychologist E. B. Titchener (1867–1927), who later came to the United States as professor of psychology at Cornell University, developed what he called *structuralism* from Wundt's analytic method. Through patient, systematic introspection and analysis of introspection, the structuralists attempted to formulate general theories of perception, thought, and consciousness.

Psychoanalysis

We have already touched on psychoanalysis in Chapter 6 and will survey it further in Chapter 15. It fits into the history of late-nineteenth-century psychology, in part as a reaction against the structuralist focus, which was exclusively on *conscious* experience. The Viennese physician Josef Breuer (1842–1925) first described what he called the "unconscious mind" in 1880, and he and his colleague Sigmund Freud developed the theory of the unconscious in their 1893 work, *The Psychic Mechanism of Hysterical Phenomena,* which laid the groundwork for psychoanalysis. Though often modified and revised, psychoanalysis has figured as the single most influential theory in modern psychology.

The Twentieth Century

Behaviorism

Another reaction against the structuralists was introduced in 1913 by John B. Watson (1878–1958), an American psychologist. With E. L. Thorndike, Watson was the leading proponent of *behaviorism,* which held that observable (*overt*) behavior, not inner experience (*covert* behavior), offered the only psychological phenomena capable of being studied; therefore, in their view, the structuralist idea of introspection was of little value. This concentration on observable events was a reaction against the structuralists' emphasis on introspection. The behaviorists focused on connections between observable behavior and stimuli from the environment, and they believed that the environment played an extremely important role in shaping behavior.

The behaviorists were profoundly influenced by the work of the Russian physiologist Ivan Pavlov (Chapter 12), who studied the relationship between stimulus and response in dogs—in particular *conditioning,* which is the learning process by which a response becomes associated with a new stimulus. Watson and his followers sought to extend Pavlov's work with conditioning to modifying human behavior.

While psychoanalysis and variations of it largely supplanted behaviorism in the 1920s and 1930s, the American psychologist B. F. Skinner did much to reawaken interest in the potential of behaviorism, even writing a speculative book, *Walden Two* (1948), which described how conditioning might be used to create a utopian society.

Gestalt Psychology

Gestalt psychology (Chapter 9) emerged about the same time as behaviorism and, like behaviorism, was a reaction against structuralism. Whereas the structuralists held that the nature of mind could be studied by analyzing the elements of consciousness, Gestalt psychologists argued that the whole was greater than the sum of the parts; that is, human beings (and other animals) perceive the world as an organized pattern, not as individual sensations or stimuli. Gestaltists sought to study behavior as an organized pattern, not as discrete instances of stimulus and response. The German psychologist Max Wertheimer (1880–1943) founded the Gestalt movement in 1912, though it wasn't generally introduced to America until the 1930s and has been most important in the study of perception and sensation.

Humanistic Psychology

In the 1950s, a group of influential psychologists sought an alternative to psychoanalysis on the one hand and behaviorism on the other. They rejected the notion that human beings were controlled by unconscious drives and motives (as the psychoanalysts held) or by the stimuli generated by the environment (the behaviorist position). The *humanists* sought to study how individuals truly *control* their own behavior and choices. The thrust of most humanistic psychology is therapeutic, with the goal of helping people to fulfill their own unique potential (self-actualization). Abraham H. Maslow and Carl R. Rogers were the chief architects of humanism.

Cognitive Psychology

Even more recently, many psychologists have become interested in such mental processes as thinking, reasoning, and self-awareness. Like the humanists, these *cognitive* psychologists see more to human personality than unconscious drives and stimulus-response. Cognitive psychology investigates how people gather and process information, then plan responses to that information. Cognitive psychology began to come into its own during the 1960s and, more recently, has been augmented by developments in computer science and information-processing theory.

The Eclectic Approach

Today, the broad field of psychology has its share of controversy and partisanship, with advocates of various approaches. However, most contemporary psychologists take an eclectic approach to the discipline, using what they consider the best and most useful features from a variety of approaches and applying them as seems appropriate in experimental as well as clinical contexts. Contemporary psychology is no longer dominated by any one particular school, but is *informed* by many of them.

CHAPTER 8

RESEARCH METHODS

CHAPTER 7 MENTIONED THE TRANSFORMATION OF PSYCHOLOGY FROM A BRANCH OF philosophy to a branch of science. The biggest single component in this transformation was the application of the scientific method (see Chapter 6) to psychological issues.

The scientific method requires a testable, falsifiable hypothesis and data to relate to that hypothesis. Psychologists acquire their data through one or more of the following broad approaches:

- By experimental methods
- By nonexperimental methods
- By correlation

Experimental Methods

Boiled down to its essentials, a psychology experiment is a situation in which something is done to subjects (this "something" is called the *experimental treatment*) in the expectation that they will be affected by it—usually in terms of a measurable change in behavior. The *subjects* of an experiment may be laboratory animals or human beings. They may be groups or individuals. *Experimental design* varies widely. For example:

- Various groups of subjects may be subjected to different combinations of experimental treatments.

- Two groups may experience variations of a single treatment.
- An individual subject may receive a variety or series of treatments.

However experiments may vary in design, they all share the following elements:

- A *hypothesis:* What the researcher predicts will happen.
- An *independent variable:* What the researcher is studying and how the groups are treated differently.
- A *dependent variable:* The behavior that is measured to determine the effect of the independent variable.
- An *experimental group:* The group exposed to the independent variable.
- A *control group:* A group not exposed to the independent variable; this group serves as a baseline against which the behavior of the experimental group can be compared.

The goal of any experiment is to understand a cause-and-effect or correlational relationship.

Operational Definitions

Most psychological experiments have as their objective supporting a general hypothesis. For example, researchers may wish to design an experiment to support their hypothesis that watching televised violence increases aggressive behavior in children. It would be impossible for these researchers to gather subject groups representative of absolutely every type of child, and it would be nearly impossible (certainly impractical) for the researchers to expose their subject groups to every conceivable example of televised violence.

Instead of attempting the impossible, the researchers use *operational definitions* to translate broad, real-world issues into what researchers can actually, feasibly do in an experiment. The independent variable (in the TV violence experiment) may be exposure to televised violence operationally defined as watching one superhero cartoon, one cop show, and one professional wrestling match. The dependent variable may be the level of aggression *operationally defined* as how many times the test subject punches a punching bag he or she is presented with.

Equivalence of Procedures and Groups

The ideal experiment measures the effect of a single independent variable. All other factors in the experiment should remain equal and unchanged from one subject to the next. This is called *equivalence of procedures.* Experimental subjects exposed to the same independent variable are treated the same in all respects. Experimental and control groups are likewise treated the same in all respects, save for the independent variable.

Failure to observe equivalence of procedures results in *confounding*, a situation in which something other than the independent variable may be responsible for differences in the behavior of the groups. For example, suppose one group of children watches the violent television shows before lunch, and another group after lunch. The researcher finds that the children in the prelunch group punch the bag more often than those in the postlunch group. Could it be that the children in the prelunch group were more aggressive because they were hungry? This extraneous variable has *confounded* the results of the experiment. The experimental circumstances should be identical for all the groups.

Similarly, subject groups should be as similar to one another as possible. Remember, the object of our TV violence experiment is to measure the effect of televised violence on the level of aggression in children. Let's say that our researcher exposes two groups to the same programs; however, the first group consists mainly of boys, and the second group mainly of girls. The researcher finds that the children in the first group hit the punching bag more frequently than those in the second group. The obvious conclusion is that boys tend to be more aggressive than girls. But this is not the kind of information the researcher is looking for. The object is to measure the effect of televised violence on the level of aggression in children. The researcher's failure to ensure *equivalence of groups* has confounded the results of the experiment.

Typically, researchers will use *random assignment* to groups to maximize equivalence of groups.

Blinds

All physicians are familiar with the placebo effect, the apparent cure produced by administering an inert sugar pill (the placebo) the patient *believes* is a miracle drug. The patient behaves according to his or her own expectations. Expecting to be cured, the patient feels better—even though the placebo contains no medicine. Experimental subjects also can be influenced by what they expect to happen; therefore, experimental subjects must be as uninformed as possible about what's happening in the experiment. This is called a *blind*. In our example, the researcher (ideally) would tell the children nothing about the programs they are about to watch or why they are being asked to watch them.

In a *double-blind* experiment, neither the subjects nor the researchers are given any information that might bias or confound the results.

One problem with blinds is a potential conflict with *experimental ethics*. Academic and professional standards prescribe ethical standards that researchers must observe in working experimentally with animals as well as human beings.

These standards typically involve obtaining the subject's *informed consent* before the subject participates in the experiment. In the TV violence experiment, for example, informed *parental* consent would be obtained, but the researcher might also advise the subjects (the children themselves) that they would be seeing some television shows that might upset them.

Statistical Analysis

The results of virtually all psychology experiments must be subjected to appropriate statistical tests to determine if the observed differences between experimental and control groups are *statistically significant*—that is, real or merely the result of chance or coincidence. In the TV violence experiment, the researcher would need to determine the average levels of aggressiveness of the control and experimental groups before and after the treatment. Using statistical methods, the researcher would then have to determine whether differences between the two groups were statistically significant. If the differences fell below the level of statistical significance, the researcher would be unable to conclude that the independent variable caused any observed difference.

Most introductory psychology courses include at least a brief introduction to statistical methods. For advanced psychology students and professional psychologists, statistics is an important tool.

Nonexperimental Research

In many situations a controlled experiment is impractical or impossible. For example, a researcher studying the effect of parents' substance abuse on substance abuse in their children could not set up an experiment in which illicit drugs were given to parents in order to see what effect this would eventually have on children. A researcher seeking such information would not perform an experiment, but would use a nonexperimental method:

- The researcher might compile and analyze *case studies* and *biographies* of drug-abusing parents and their children.
- The researcher might *survey* these groups, either interviewing them or using questionnaires.

While nonexperimental methods greatly broaden the scope of psychological inquiry, these methods have serious shortcomings.

- It is impossible to know all the factors that may be relevant to the information obtained.
- It is difficult, perhaps impossible, to determine with certainty cause and effect, because confounding influences cannot always be identified.

- In the case of surveys, biases may be introduced. It may be difficult to get a truly representative response.
- Surveys are also prone to distortion. How do you know respondents are telling the truth? (This is a particularly serious problem, for example, with surveys concerning sexual behavior.)

Nonexperimental methods challenge the researcher to create tools, such as questionnaires, that avoid or minimize the method's inherent weaknesses.

A third nonexperimental method is the *quasi-experiment.* In this case, the researcher identifies subjects and conditions as they naturally occur, then observes what happens. For example, a researcher may want to assess how spanking by parents affects aggression in children. Using survey methods, the researcher may identify two groups: parents who spank and parents who do not spank. Over a period of years, the researcher may observe the children of these two groups of parents, periodically interviewing them and asking questions that measure aggression (for example, "How many times have you been sent to 'dentention' after school?"). At the end of the study, results are tabulated and analyzed.

The type of study or quasi-experiment just described is a *longitudinal design,* because it follows subjects over a period of time. Another possible design is *cross-sectional design,* which also assesses change, but does not follow groups over time. In the spanking study, the researcher might interview a cross section of children divided by age: four-year-olds who have been spanked versus four-year-olds who have not been spanked, six-year-olds who have been spanked versus six-year-olds who have not been, and so on, maybe well up into the teen years, depending on the scope of the quasi-experiment.

More About Statistics: Correlation

Simple conclusions about cause and effect cannot always be made and are not always desirable. For example, it would be difficult to imagine the spanking researcher concluding definitively that spanking *causes* children to be aggressive. However, the researcher might use *correlation* to show the statistical relationship between spanking by parents and the level of aggression in their children.

The *coefficient of correlation* is a mathematical index of the correspondence between two measures. The coefficient can vary from 0 to +1.00 or from 0 to -1.00. The first is the range of *positive correlation*; the second is the range of *negative correlation.*

- *Positive correlation:* If two measures vary in the same direction across subjects, they are said to be positively correlated. Let's say our spanking researcher discovers that spanked children are consistently more aggressive than children whose parents do not spank them. The researcher may find a positive correlation between the frequency of spankings and the frequency of aggressive episodes. The researcher may further conclude that a positive correlation exists between spanking and aggression.

A correlation of 0 means that no correspondence exists. Correlations near 0 are weak, whereas correlations between .20 and .60 are moderate. Correlations greater than .60 are considered strong.

- *Negative correlation:* If two measures vary in opposite directions, they are said to be negatively correlated. Let's say the spanking researcher encounters a startling result: The more often children are spanked, the less often they are involved in aggressive episodes. If this were the case, there would be a negative correlation between frequency of spanking and the number of aggressive episodes in which the child was involved.

Correlation can provide important clues about cause-and-effect relationships, but, in and of itself, correlation does not *prove* cause and effect. Psychologists learn to interpret correlations cautiously and conservatively. This caution and conservatism, taking correlations as clues rather than as definitive proofs, separates the informed and seasoned researcher from the overeager novice.

CHAPTER 9

PHYSIOLOGICAL PSYCHOLOGY

YOU'RE STUDYING CELLS, GENETICS, THE ORGANS OF SENSATION, CHEMICAL TRANSmission of nerve impulses . . .wait a minute! Have you signed up for Intro to Psych or Intro to Biology?

It does come as a surprise to some students in introductory psychology courses when they find a portion of the course remote from such subjects as emotions, motivation, and thought: physiology, the *biological* study of the functions of organisms and their parts. What is *this* material doing in a survey of psychology?

For one thing, we have already seen that psychology is an extraordinarily diverse field. Physiological issues are part of that diversity. More important, however, psychology is a science, and, in science, all behavior is measurable. Moreover, it is a cornerstone of modern psychology that all behavior is assumed to have a physical cause. That which is physical can be observed and measured—whether it lies within the organism or outside of it.

This is the rationale for including *physiological* psychology within the field of psychology; however, the subject may still strike you as somehow out of place alongside discussions of such subjects as psychoanalysis and complex issues of how learning takes place. The reason is that there is a great gap of knowledge between what is understood about the physiological basics of behavior and what is understood about behavior from observing its external manifestations and results. Put another way, it is one thing to describe the physical and chemical processes associated with the transmission of a nerve impulse, but it is quite another to describe just how such processes manifest themselves as what we call *thought*. The links between *overt* behavior (as manifested in electrochemical processes in nerve cells, for example) and *covert* behavior (thought, mental processes) have yet to be explained. The prospect that these links will one day be explained is one of the most exciting frontiers of psychology.

Nerve Cells and Nervous System

The most basic structures of interest to the physiological psychologist are *neurons*, the specialized cells that transmit and process information in the form of electrochemical impulses. The human body has some 12 billion neurons. Each neuron consists of a cell body, from which one or more short extensions, *dendrites*, and one long extension, the *axon*, project. Axons are sheathed in a fatty envelope of material called *myelin*. Bundles of myelin-sheathed axons function as *nerves*.

Animals, including human beings, have three types of neurons:

1. *Sensory neurons* receive stimuli from the external environment.
2. *Motor neurons* transmit impulses from the brain and spinal cord to muscles or glands, stimulating contraction or secretion.
3. *Interneurons* connect sensory and motor neurons and also carry stimuli in the brain and spinal cord.

Neurons propagate nerve impulses. Such an impulse is an electrochemical phenomenon. In an inactive neuron, a neuron at rest, the cell fluid, or *cytoplasm*, is negatively charged with respect to the outside of the cell. When this difference in electrical charge disappears, a nerve impulse is generated. This happens when a stimulus contacts the tip of a dendrite, increasing the permeability of the cell membrane to sodium ions. (An *ion* is an atom or group of atoms that has gained a net electrical charge by gaining or losing one or more electrons. The sodium ions have a net *positive* charge.) When the cell membrane becomes more permeable, the positively charged sodium ions rush back into the negatively charged cytoplasm, thereby removing the difference in electrical charges and creating an electrochemical pulse—the *nerve impulse*.

The nerve impulse travels through the dendrite, then through the cell body, and, finally, down the axon. At the end of the axon is a fluid-filled space called a *synapse*. Synapses may occur between the axon of one neuron and the dendrite of another or between neurons and muscle fibers, forming a *neuromuscular junction*. When the impulse reaches the end of the axon, it causes the release of chemicals called *neurotransmitters*. Molecules of these substances accumulate at the synapse, increasing the membrane permeability of the next dendrite. This causes a new impulse to be generated in that neuron. (Again, positively charged sodium ions rush into the negatively charged cytoplasm of the neuron.) Thus the impulse is transmitted from neuron to neuron.

Nerve impulse transmission either occurs or does not occur. It is a binary, on-or-off situation. The degree or strength of the total impulse that passes through the bundle of neurons that make up a nerve is determined not by the intensity of a particular impulse in a particular neuron, but by the total number of neurons that are stimulated.

CNS and PNS

In human beings, nervous coordination is carried out by two nervous systems, the *central nervous system* (CNS) and the *peripheral nervous system* (PNS). The nerves of the CNS lie within the brain and spinal cord, while those of the PNS extend outside of these structures to and from the CNS.

The *spinal cord* extends from the base of the *brain* to the bottom of the backbone. The spinal cord detects certain stimuli and responds to them directly (without reference to the brain) in what is called a *reflex arc.* (The familiar knee jerk that results when your doctor tests your reflexes by tapping your knee with a rubber mallet is an example of a spinal reflex arc.) The spinal cord also serves as a kind of neural trunk line, distributing nerves from the brain to the rest of the body.

The brain is the most complex of all human organs. It is a complex mass of nervous tissue and is the site of consciousness, sensation, memory, and intelligence. In general, the superficial portion of the brain, the *cerebral cortex,* controls the higher intellectual functions, while the structures of the inner portion control the body's many autonomic activities. We will survey this organ further in just a moment.

The peripheral nervous system consists of the *sensory somatic system,* which carries impulses from the external environment and the senses, and the *autonomic nervous system,* which involuntarily prepares the body to meet various threats by increasing the heartbeat, constricting arteries, and dilating pupils. The PNS is called *peripheral* not because it is somehow less important than the CNS, but because its nerves all lie outside of (are peripheral to) the *central* nervous core, the brain and spinal column.

The autonomic nervous system is further subdivided into the *sympathetic nervous system* and the *parasympathetic nervous system.*

The sympathetic nervous system prepares the body to respond to emergency situations by increasing heart rate and respiration while simultaneously shutting down (at least partially) digestion and other body processes that aren't immediately necessary in an emergency. The sympathetic nervous system also stimulates certain glands of the *endocrine system* to secrete substances that temporarily increase physical strength (adrenaline is one such substance) and that temporarily reduce the perception of pain (these substances are called *endorphins*). Thus the sympathetic nervous system helps to prepare the body to cope with stress and to survive the immediate threat.

The parasympathetic nervous system comes into play after the emergency has passed. It operates to return the body systems to normal functioning.

The sympathetic and parasympathetic nervous systems also play important roles in sexual arousal and orgasm. Arousal is chiefly a function of the parasympathetic system, whereas orgasm is a function of the sympathetic system.

The operation of the sympathetic and parasympathetic nervous systems are important aspects of the *fight-or-flight syndrome.* That is, they operate in threatening situations to prepare the organism more effectively to fight or to flee. (Various emotional disorders, especially *anxiety disorders,* involve overactivity of the sympathetic nervous system, which may respond unrealistically, even in the absence of external threat.)

The Senses

Besides the brain, spinal cord, and nerves, the *senses* are the other key organs in the nervous system. They are discussed later in the chapter.

The Brain

The greater part of whatever portion of your introductory course is devoted to physiological psychology will doubtless focus on the brain, the center of the organism's information processing and, in human beings, the seat of what we call *mind.*

Mapping the Brain

Advances in experimental techniques with animals and surgical techniques with human beings have allowed physiological psychologists to identify the areas and structures of the brain that are associated with specific neurological functions. The mapping is far from complete, and it is also a little misleading, since, according to the *principle of mass action,* most behaviors involve the entire brain, not just this or that part. Nevertheless, it is highly useful to survey the principal brain structures and areas. On the grossest level, the brain may be divided into *hindbrain, midbrain,* and *forebrain.* Together, the hindbrain and midbrain form the *brain stem,* the part of the brain that is located atop the spinal cord. The forebrain is the large mass that occupies the cranial cavity.

- In evolutionary terms, the hindbrain is the most primitive of the brain's structures. It coordinates many of the body's involuntary functions, such as heart rate and breathing. Within the compass of the hindbrain is the *cerebellum,* which is involved in balance and in gross motor coordination.
- The midbrain represents a step up in evolution and plays a role in controlling certain sensory functions, including the transmission of visual information from higher parts of the brain to the cerebellum. Within the midbrain is the *reticular formation,* which plays a role in regulating sleep and arousal, as well as attention.

- The forebrain is the most advanced structure of the brain and the largest.

Forebrain Structures

Four major structures comprise the forebrain:

- The *thalamus,* located where the brain stem and reticular formation end, plays a part in sleep and arousal and in relaying sensory information to the *cerebrum.*
- The *hypothalamus* is a small structure that is involved in hunger, thirst, sex, stress reactions, and emotion.
- The *limbic system* is closely related to the hypothalamus and is involved in regulating eating, drinking, and sexual behavior. The limbic system also apparently contributes to the regulation of aggression. Within the limbic system, the *hippocampus* functions in memory, particularly short-term memory.
- The *cerebrum* is the largest single structure in the brain and is the seat of higher brain functions, including what we characterize as *mind.*

Cerebral Structures

The cerebrum consists of an inner layer and an outer *cerebral cortex.* It is this cortex, a convoluted structure, that is the site of higher brain activity.

The cortex is divided into two *hemispheres,* connected by an inner structure called the *corpus callosum.* Each hemisphere is further divided into four lobes: the frontal, temporal, parietal, and occipital lobes, corresponding to the front of the brain, the temple area of the brain, the sides, and the rear areas, respectively. Much of the cortical area has been mapped, so that different portions of the cortex have been identified with different sensory and somatic functions. Some of the major cortical areas include the following:

- The *motor cortex,* along the rear of the frontal lobe, is involved in bodily movements.
- The *somatosensory cortex,* along the front of the parietal lobe, is involved in bodily sensation.
- The *auditory cortex,* in the upper temporal lobe, is involved in hearing.
- The *visual cortex,* occupying most of occipital lobe, is involved in vision.
- *Association areas,* located in large areas of the frontal, temporal, and parietal lobes, are the sites of all that we call thought. These areas have been sufficiently studied to produce such conclusions as the location of speech functions in *Broca's area* and elsewhere in the left hemisphere. The associative areas on this side of the brain also seem to be most involved in logical thinking and in mathematics. Corresponding areas on the right hemisphere tend to

be involved in pattern-oriented thought, including art and music. Some have suggested that as the left brain is oriented toward logical thinking, the right brain is oriented toward more imaginative and creative thought.

Sensation and Perception

Closely related to study of the physiological aspects of psychology are the processes of sensation and perception.

- *Sensation* encompasses the means by which we obtain information about the external environment. *Sensations* are the effects on the nervous system that occur when specific forms of energy act upon and stimulate our sense organs.
- *Perception* is the higher processing of sensory information after it reaches the CNS.

Sensation is essentially a physiological process, whereas perception draws on higher mental functioning—cognition, which involves recognizing, filtering, interpreting, and organizing information based, in part, on experience and learning.

Measuring Sensation

Sensation is, of course, a very familiar phenomenon, yet, precisely because it is so familiar—literally, so *natural*—it presents a challenge to the researcher who would quantify sensation in order to measure it. Classical measurement methods use *sensory thresholds:*

- An *absolute threshold* is the minimum stimulus required to produce a sensation. Anyone who has had their hearing tested by a hearing specialist is familiar with this concept.
- A *difference threshold* is the minimum degree to which a stimulus must change in order to be perceived as having changed. The *just noticeable difference* (JND) is the point at which the difference is recognized correctly 50 percent of the time. If you have agonized over precisely what shade of peach to paint your living-room walls, you have probably experienced a difference threshold as you try to distinguish between two very similar hues.

The nineteenth-century psychologist Ernst Weber formulated *Weber's law,* which states that the JND varies as a function of the intensity of the stimulus. The more intense a stimulus is, the greater the JND must be.

By defining and using thresholds (also taking into account *sensory adaptation,* by which sensory sensitivity tends to decrease with repeated exposure to the same or similar stimuli), researchers can arrive at reasonably useful measurement of sensation.

Visual Sensation

Human beings are highly visual organisms, and electromagnetic energy in the visible range is relatively easy to measure, as are the sensory thresholds for visual stimuli. It is no wonder, then, that vision has been more thoroughly studied than any other sense.

Light is electromagnetic energy. It enters the eye through the *cornea* (a protective outer surface) and is focused by the *lens* onto the *retina,* a field of specialized neurons called *rods* and *cones.* (The amount of light that is sent through the lens to the retina is regulated by the *iris,* which enlarges and contracts the *pupil,* analogous to the variable aperture of a camera.)

The structure within the eye that is of greatest interest to physiological psychologists is the retina, over which are distributed some 100 million rods and 6 million cones. The rods are sensitive to low levels of light, but can "see" only black and white and shades of gray. The cones, most of which are concentrated at the *fovea,* near the center of the retina, respond to color and are capable of discerning finer detail, but require higher levels of light. The rods and cones feed into *bipolar cells,* which relay the visual impulses to some one million *ganglion cells* on the optic nerve, from which the impulses travel up the optic nerve on their way to the occipital lobes of the brain. Presumably, a certain amount of visual processing takes place before the impulses reach the brain, probably in the ganglion cells.

The optic nerves merge and partially cross at a structure called the *optic chiasma.* The effect of this merger and crossover is that visual sensations from the right visual field (that is, the left half of each retina) are sent to the left occipital lobe, and sensations from the left visual field go to the right occipital lobe. The process by which the visual field processed by the occipital lobe is divided is called *visual mapping.*

Most of the pioneering work in visual perception—how the brain makes visual sensation intelligible—was done by the Gestalt psychologists early in the twentieth century. The Gestaltists studied:

- *Figure-ground relationships*: How we perceive foreground images as distinct from a background.
- *Closure:* Our tendency to see figures as complete—even when portions are missing.
- *Proximity:* Our tendency to group objects perceptually, based on their proximity to one another.
- *Similarity:* Our tendency to group items according to similarities among them.
- *Continuity:* The tendency, when objects are arranged in certain patterns, to perceive the patterns rather than the individual objects.

Another interesting area of study, which has interested more recent psychologists, is the perception of depth and distance, particularly the relationship between *binocular cues* (perception based on information gathered from both eyes) and *monocular cues* (perception based on information from one eye). Binocular vision creates retinal disparity, a slightly different view in the retina of each eye. The closer an object, the greater the *retinal disparity.* The perception of depth and distance created by binocular vision is aided by monocular cues, including size (near objects appear larger than far objects), interposition (nearer objects block farther objects), height in field (farther objects appear higher in the field of vision than nearer objects), linear perspective (parallel lines appear to converge at a point on the horizon—the "vanishing point"), and texture gradients (patterns of texture become finer with distance).

In addition to depth and distance perception, psychologists have also been interested in how we create *perceptual constancies:*

- *Size constancy:* If we move closer to on object, it looms larger in our field of vision, yet we do not *perceive* its size to have increased.
- *Brightness and color constancy:* Although the apparent brightness and color of an object changes under different levels of illumination, we do not perceive the object as having changed.
- *Shape constancy:* The apparent shapes of objects change as we change our position and point of view, yet we perceive the shape of the objects as constant.

The phenomenon of color vision has drawn the interest of scientists since the nineteenth century. The first modern theory of color vision was the *trichromatic theory,* which identified three different types of cones on the retina: one sensitive to blue light, another to green-yellow, and a third type sensitive to red. The perception of color, according to this theory, is the result of different combinations of "firing" among these three cones. Later, the *opponent-color theory* elaborated on the trichromatic theory. At the level of the receptors (the cones), the trichromatic theory holds true, but at the level of the bipolar cells or beyond, two additional systems process color vision: a red-green system and a blue-yellow system. Each of these systems assesses the information from the firing of the cones, then sends the processed information to the brain.

Auditory Perception

The psychology of *audition* (hearing) has been researched somewhat less thoroughly than that of vision, but it has been extensively studied nevertheless. The organ of hearing is, of course, the ear, which transforms sound waves into nerve impulses by funneling these waves to the *eardrum,* which vibrates against three tiny bones, the *malleus* (hammer), *incus* (anvil), and *stapes* (stirrup). These bones, in turn, transmit the vibrations to the *cochlea.* Fluid within this structure vibrates in response to vibrations transmitted from the three bones. Within the cochlea, these vibrations are picked up by hair cells (of the *organ of Corti*) that excite nerve impulses, which are carried along the *auditory nerve* to the brain.

According to the *frequency theory,* the frequency at which the hair cells of the organ of Corti send neural impulses to the brain matches the frequency of the sound entering the ear. When it was discovered that neurons cannot fire faster than about a thousand times a second (too slow even for sounds in the middle frequency range, let alone those at higher frequencies), E. G. Wever augmented the frequency theory with the *volley principle,* arguing that neurons work together in sequenced loops to achieve higher effective firing rates.

Taste and Smell

These related senses operate through *chemoreceptors,* specialized cells that are stimulated variously by the molecules and ions of substances in the environment. When stimulated, nerve impulses are generated and transmitted via the *olfactory nerve* to the brain. Psychologists have performed experiments to investigate how information from the receptor organs called *taste buds* interacts with information from the olfactory receptors to produce what we perceive as flavor.

Touch and Other Senses

Receptors for touch and pain are called *Pacinian corpuscles* and are located in the skin, in muscles, and in tendons. Changes in pressure cause the generation of nerve impulses. The sense of balance is largely centered in the inner ear, where changes in the orientation of fluid in the *semicircular canals* stimulate hair cells to fire certain nerves. Internal or visceral senses (such as fullness of bladder or bowel) are created by *stretch receptors* in the muscles and by *carbon dioxide receptors* in the arteries. *Kinesthesis* involves internal receptors that tell our brains what parts of our bodies are doing. For example, we don't have to look at our upraised arm to know that it is upraised.

Endocrine System

Ultimately, the nervous system and the other systems in human beings and other animals function to establish and maintain *homeostasis*—a balance between internal processes and external conditions. In addition to the nervous system, your introductory psychology course will probably also survey the human *endocrine system.* As the nervous system functions to achieve and maintain the body's nervous coordination, so the endocrine system functions to achieve and maintain *chemical coordination.*

Animal bodies, including the human body, use a system of *endocrine glands* to secrete *hormones*, proteins and lipids that bring about changes in the body and help coordinate body systems. The endocrine glands secrete their various hormones directly into the bloodstream, which carries the hormones to *target organs*—the organs the particular hormones are supposed to stimulate. In contrast to such ducted body glands as the salivary glands (which secrete saliva for digestion) and the sweat glands, the endocrine glands have no ducts and are sometimes referred to as the *ductless glands*. (The ducted glands are sometimes called *excorine glands.*)

Pituitary Gland

In human beings, the *pituitary gland* is located at the base of the brain. It secretes important hormones that play key roles in regulating growth and maturation. One of these hormones, *human growth hormone* (HGH), accelerates protein synthesis to promote body growth. Oversecretion of the hormone produces gigantism, while undersecretion produces dwarfism. *Lactogenic hormone* (LH), also called *prolactin*, promotes breast development in females and stimulates the secretion of milk.

The endocrine glands often work in an interrelated fashion. The pituitary produces *thyroid stimulating hormone* (TSH), which controls secretions from the thyroid gland. *Adrenocorticotrophic hormone* (ACTH) controls secretions from the adrenal gland.

The pituitary also secretes:

- *Follicle stimulating hormone* (FSH): In females, this stimulates development of a follicle, which contains an egg cell. In males, it stimulates sperm production.
- *Luteinizing hormone* (LH): In females, LH completes the maturation of the follicle and also stimulates formation of a structure called the *corpus luteum*, which secretes more female hormones.
- *Interstitial cell stimulating hormone* (ICSH): The male counterpart of LH, it stimulates the secretion of male hormones in the testes.
- *Melanocyte stimulating hormone* (MSH): Stimulates production of *melanin*, the skin pigment.
- *Antidiuretic hormone* (ADH or *vasopressin*): Produced in a part of the brain called the *hypothalamus*, this hormone is stored and released by the pituitary. It stimulates water reabsorption by the kidneys.
- *Oxytocin:* Another hormone produced by the hypothalamus and stored and released by the pituitary, oxytocin stimulates contractions of the uterus during the birth process.

Because its hormones regulate so many functions, including the work of other glands, the pituitary is sometimes referred to as the "master gland."

Thyroid Gland

Located (in human beings) against the pharynx, at the base of the neck, the *thyroid gland* produces *thyroxine* and *calcitonin.* Thyroxine regulates the body's metabolic rate. Calcitonin regulates the level of calcium in the blood.

Parathyroid Glands

These structures are found on the posterior (rear) surfaces of the thyroid gland and produce *parathyroid hormone,* or *parathormone.* This substance regulates calcium metabolism.

Adrenal Glands

Located on top of the kidneys, the *adrenal glands* consist of an outer *cortex* and an inner *medulla.* The cortex secretes several substances collectively called *corticosteroids.* These stimulate and regulate a variety of body functions, including mineral metabolism, glucose metabolism, and protein synthesis.

The medulla portion of the adrenal glands secretes *epinephrine* (also called *adrenaline*) and *norepinephrine* (noradrenaline). Epinephrine increases the heart rate and blood pressure, thereby increasing the blood supply to the skeletal muscles. In times of fear or excitement, when the individual is threatened, this "adrenalin rush" helps the body prepare for what is frequently called the *fight-or-flight response.* Increased blood supply to the skeletal muscles enables more effective defense in the form of fighting or fleeing. Norepinephrine serves to augment the effects of epinephrine.

Pancreas

The human *pancreas* is located behind the stomach. Cell clusters called the *islets of Langerhans* secrete two important endocrine hormones, *insulin* and *glucagon.* Insulin promotes passage of glucose molecules into the body and regulates their metabolism. Without insulin, glucose is simply removed from the blood and excreted, resulting in diabetes mellitus, a disease characterized by extremely sluggish body metabolism.

Glucagon stimulates the breakdown of glycogen to glucose in the liver. It also releases fat from *adipose tissue* (tissue where fat is stored) so that it can be used for the synthesis of carbohydrates.

Ovaries and Testes

In females, the *ovaries* secrete *estrogens,* which promote development of secondary female characteristics. In males, the *testes* secrete *androgens,* which promote secondary male characteristics.

Pineal Gland

Located in the midbrain, this small gland is something of a mystery, though it seems to play a role in regulating mating behavior and in regulating the body's day-night cycles.

Thymus Gland

Found in the neck tissues, the *thymus* secretes *thymosins,* which act on the immune system to stimulate T-lymphocytes.

Other Endocrine Secretions

Various body tissues secrete *prostaglandins,* which have diverse effects. Various prostaglandins stimulate smooth-muscle contraction, lower or raise blood pressure, decrease and increase the clotting ability of blood, enhance ion transport across some membranes, stimulate inflammation, and inhibit the breakdown of fat in adipose tissue.

Kidney cells produce *erythropoietin,* which plays a role in the production of red blood cells, and digestive glands secrete *gastrin* and *secretin,* which aid in regulating the digestive process.

Genetics

One other physical aspect of life that is often addressed in introductory psychology courses is heredity and genetics. The defining property of a species is that members share a set of inherited characteristics. Yet, within a species, there may be wide variation. Individuals inherit characteristics that are common to the species, as well as characteristics in varying degrees unique to the individual. The inherited characteristics—called *traits*—of an organism (human beings included) are transmitted to that organism by *genes.*

While the science of genetics advanced greatly beginning in the 1950s and 1960s with the discovery of how the molecule DNA operates as the central molecular mechanism of genetic inheritance, the study of genetics began in the 1860s and 1870s with the work of Gregor Mendel (1822–1884), a humble Austrian monk. Mendel experimented with pea plants, focusing on patterns of inheritance of seven obviously varied and visible traits: plant height, pod shape, pod color, seed shape, seed color, flower color, and the location of the flower on the plant. Mendel chose well in focusing on pea plants, because they are self-pollinating, developing individuals that are *homozygous* for particular characteristics.

The different forms of a gene are called *alleles.* A pea plant (for example) may have an allele for tall plants and one for short plants. A tall *homozygous* pea plant will have two tall alleles. Its line of inheritance for this characteristic is said to be *pure.*

Mendel took various *pure-line* pea plants and artificially cross-pollinated them with other pure-line plants, then noted the resulting offspring.

After years of patient experimentation, Mendel formulated three cornerstone laws of inheritance:

1. *Mendel's law of dominance:* When an organism has two *different* alleles for a trait, one allele dominates (and is called a *dominant trait*). Nondominant alleles are called *recessive traits.*
2. *Mendel's law of segregation:* During gamete formation by a diploid organism, the pair of alleles for a given trait *segregate* (separate)—as in meiosis. If hybrid (*heterozygous*) individuals are crossed, the recessive trait segregates out at a ratio of 3 to 1; that is, for every three offspring exhibiting the dominant trait, one will exhibit the recessive trait.
3. Mendel's law of independent assortment: During meiosis, the members of a gene pair separate from one another independently of the members of other gene pairs. That is, each trait is inherited independently of others.

Mendel spoke of "factors," not genes. He neither studied nor speculated on the actual mechanisms of inheritance; he only observed—and brilliantly so—the outward workings of inheritance. In 1911, the biologist Thomas Hunt Morgan (1866–1945) began to work with fruit flies to study the mechanisms of inheritance more closely. He concluded that the *chromosome* is the means by which hereditary traits are passed from one generation to the next.

Sex cells (eggs and sperm) replicate through meiosis. During this process, threadlike cellular structures called *chromosomes* duplicate and pass into two cells. The chromosomes of the two cells then separate and pass into four daughter cells. The parent cell has two sets of chromosomes (so is called *diploid*), while the daughter cells each have a single set of chromosomes (so are called *haploid*). These daughter cells are the *gametes,* the sex cells that will fertilize or be fertilized by the gametes of another organism (of the same species), creating a diploid *zygote,* which will develop into an *embryo* and, ultimately, into a new organism. The new organism combines chromosome-borne traits of both of its parents.

Through his work with fruit flies, Morgan further concluded that chromosomes are composed of smaller, discrete units he called *genes.* It is the genes that actually carry specific traits. The genes move with the chromosomes during the process of cell division. (Morgan also proposed that, if the genes change—*mutate*—the traits they control change.)

Predicting Genetic Possibilities

In some introductory psychology surveys, you may be introduced to the *Punnett square,* which is used to predict the *genotypic* and *phenotypic* ratios that result from the combination of various gametes. The *genotype* of an organism is all the genetic traits of the organism, as revealed by breeding experiments. The *phenotype* is the visible traits of the organism—the visible *expression* of genotype—which can be seen in the organism without any breeding experiments.

Here is a simple version of one of Morgan's fruit fly experiments. Some fruit flies exhibit long wings. Others exhibit short, vestigial wings. The trait for long wings is dominant—and is, therefore, expressed in shorthand form with a capital letter: L. The trait for vestigial wings is recessive and is expressed with a lowercase letter: l. A fruit fly is *heterozygous* for wing length if it bears genes for both long (L) and vestigial (l) wings. What happens if two such fruit flies mate? You can sketch a Punnett square to find out:

Male _ Female
Ll

	L	l
L	LL	Ll
l	Ll	ll

Based on the Punnett square, you may conclude that the next generation of fruit flies will consist of:

1 homozygous dominant long-winged fly (LL)
2 heterozygous dominant long-winged flies (Ll)
1 homozygous recessive short-winged fly (ll)

Note that these results list ratios of *genotype*. The *phenotypes* (visible characteristics) represented here are three long-winged flies to one short-winged fly. You cannot tell by looking at the long-winged flies whether they are homozygous for the trait or heterozygous for it.

Crossovers and Mutations

Theoretically, genes are linked on chromosomes, so they are inherited as a group on a particular chromosome. In actuality, linked groups may be broken during meiosis, and chromosomes may exchange homologous parts in a phenomenon called *crossing over.*

It is also possible that genes can be changed through *mutation.* Mutations are usually random—though they may also be caused by certain chemicals and by radiation. Mutations are usually recessive, and they are usually (but not always) harmful.

Deletion is a type of mutation caused by the loss of a piece of chromosome. *Duplication* results when a piece of a chromosome adheres to another chromosome, giving that chromosome too many genes. In the case of *inversion,* genes become rearranged on a chromosome, thereby changing the genetic sequence on the chromosome and preventing chromosomes from matching up properly during meiosis. Finally, individual genes can change due to *point mutations.*

Another type of mutation, *polypoidy,* results when cells develop extra sets of chromosomes. This condition is sometimes purposely induced with chemicals to create polypoid fruits and flowers, which are very large.

Sex-Linked Traits

In human beings, one pair of chromosomes determines the sex of the offspring. In females, this pair is designated XX, while in males it is designated XY. The Y chromosome is much shorter than the X chromosome. This means that when a gene occurs on one X chromosome in a female, a corresponding gene also probably occurs on the other X chromosome of the pair. However, in the male, such a corresponding gene probably will not be present on the short Y chromosome. For example, the trait of red-green color blindness is recessive and is carried on the X chromosome, as is the dominant gene for normal vision. If one of the female's X chromosomes has the gene for color blindness, the chances are that she will have the dominant gene for normal color vision on her other X chromosome. Thus very few females are color blind. If, however, the color blindness gene is present on a male's chromosome, the dominant gene for normal color vision will *not* be present on the Y chromosome, because it is so short. Males, therefore, have a much greater chance of inheriting the color blindness phenotype.

Traits such as color blindness and hemophilia are said to be *sex-linked,* because the genes for them occur on the X chromosome.

Action of Multiple Alleles

According to classical genetic theory, *alleles* are two or more genes that occupy the same positions on homologous chromosomes. Alleles are separated from each other during meiosis, then come together again at fertilization, when homologous alleles are paired—one from the sperm cell and one from the egg cell. Yet the inheritance of some traits cannot be explained by the action of a single pair of alleles. Sometimes multiple alleles are involved, as in the inheritance of blood type in human beings:

Blood type A may be produced by genotype I^AI^A or I^Ai

Blood type B may be produced by genotype I^BI^B or I^Bi

Blood type AB may be produced only by genotype $I^A I^B$

Blood type O may be produced only by genotype ii

Importance of Genetics to Psychology

A basic understanding of genetics is important in making assumptions and drawing conclusions about issues of nature versus nurture—whether a particular trait or behavior is likely to be the result of heredity or the influence of the environment. A number of anomalies and disorders are genetic in nature, some of which have significant effects on psychological traits. Down syndrome (mental retardation) and color blindness are just two examples.

CHAPTER 10

DEVELOPMENTAL PSYCHOLOGY

DEVELOPMENTAL PSYCHOLOGY LOOKS AT HUMAN BEINGS FROM THE POINT OF VIEW of heredity and environment and studies how they change—physically, cognitively, and in terms of personality—over a period of time. That period of time may encompass a number of years, childhood, or even conception to death.

Stages and Periods

Developmental psychologists attempt to identify developmental *milestones* common to all human beings. In the work of a number of psychologists, including Freud, Jean Piaget, and Erik Erikson, developmental stages play important roles.

A *stage* is defined by specific kinds of physical and psychological changes. A *period* is defined by a typical age range used to describe development. We will look at some stages in surveying the work of Piaget and Erikson later in this chapter (Freud's work is surveyed in Chapter 15). Here is the typical way in which psychologists divide the human life span into periods:

1. Conception
2. Prenatal period: gestation
3. Infancy: birth to 12 (or 24) months
   a. Neonatal period: birth to three weeks
4. Toddlerhood: from first steps to 24 months

5. Preschool period: 24 months to 6 years
6. Middle childhood: 6 years to puberty
7. Adolescence: 12 or 13 to sexual maturity (girls); 14 or 15 to sexual maturity (boys)
8. Adulthood: approximately 18 to 40
9. Middle adulthood: approximately 40 to 65
10. Late adulthood: 65 on

Prenatal Development and Psychology

In the portion of your course devoted to developmental psychology, you may be surprised that some time is spent discussing development from conception to birth. But the fact is that the prenatal period is part of life, and it is a part of life in which certain skills and adaptive behaviors develop. The prenatal period is also a time in which disorders and defects can develop, which will adversely affect the psychological development of the child. Important prenatal disorders include rubella (German measles) in the mother. This disease can affect the developing fetus, producing many birth defects, including blindness, deafness, and mental retardation. Fetal alcohol syndrome (FAS), caused by the mother's use of alcohol during pregnancy, can create physical and mental problems in the child. Use of various drugs during pregnancy can also have devastating physical and psychological effects. Substances such as drugs, chemicals, and even viruses that can cause problems with fetal development are called *teratogens.*

Infant Development

Development during infancy has been closely and extensively studied. Learning during this period is rapid and dramatic. We think of newborns as essentially helpless, but, in fact, they are equipped rather well for survival. We are born with *survival reflexes,* including breathing, sucking, and swallowing. Somewhat more complex is the *rooting reflex.* Gently stroke the cheek of a newborn, and he or she will turn toward the touch—in search of the breast.

Two other "primitive" reflexes may have played some evolutionary role in survival:

- The *Moro reflex* is seen when a baby hears a loud noise or suddenly finds him- or herself in a new bodily position. The newborn flings out his or her arms sideways, then brings them back in. This may be an effort to grab onto something.
- The *Babinski reflex* is a spreading and curling of the toes when the bottom of the foot is stroked.

Both the Moro and Babinski reflexes fade and disappear within the first year after birth. Their disappearance is a sign of developmental progress and normality.

For many years, psychologists, physicians, and others believed that newborns possessed very limited sensory and perceptual capabilities. More recent studies have suggested that their perceptual and sensory capacity is actually quite extensive. Of course, testing these faculties is difficult in subjects who cannot speak. Careful observation is called for.

Psychologists can use the *habituation method* or the *preference method* to measure sensory-perceptual capabilities in the newborn.

- *Habituation:* Researchers can observe the neonate's ability to distinguish differences between tones by playing a continuous tone and observing the neonate's response. Typically, the initial response is interest manifested by heightened alertness. After a time, however, the baby will appear bored, having "habituated" to the tone. If the tone is changed, even slightly, signs of interest return, indicating that the neonate is sensitive to and can discriminate between changes in tone.
- *Preferences:* Researchers can use conditioning techniques (see Chapter 12) to measure the neonate's preferences. In some experiments, a special nipple is connected to equipment that measures the rate of suckling. The researcher can associate a given stimulus with rapid suckling and another stimulus with slow suckling; that is, if the neonate suckles at a rapid rate, he or she will be exposed to one stimulus, and if the neonate suckles at a slow rate, he or she will be offered another stimulus.

Using the preference approach, researchers have found that neonates not only can distinguish human voices from other sounds, but prefer human voices to other sounds. They also prefer higher-pitched (female) voices to lower-pitched (male) voices, and they quickly learn to prefer their mother's voice to others. They also prefer human faces to other objects, and it is clear that they can distinguish their mother's milk from that of another woman.

Motor development is easier to observe and measure than sensory and perceptual development. Psychologists have established *developmental norms* for a host of motor milestones. Some of the most prominent that psychologists (and parents!) watch for include:

- Props up his or her own body: about four months
- Sits without support: six months
- Crawls (belly on floor), then creeps (belly not touching floor): 7 to 12 months
- Stands without support: about 12 months
- Walks: 12 to 18 months

While a great deal of "average" data has been compiled, it is still a fact that many perfectly normal children hit various milestones earlier or later than the average.

Two Key Developmental Theories

It requires a major intellectual leap to move from observation of development to fashioning developmental theory from observation. While a number of psychologists have offered theories of development, three figure as most influential. One is Freud's psychoanalytic view, which is discussed in Chapter 15. The other two prominent developmental theories are those of the Swiss psychologist Jean Piaget (1896–1980) and the German-born American psychologist Erik Erikson (1902–94).

Piaget's Cognitive Theory

Jean Piaget devoted a large portion of his life to observing the development of his own three children. His major interest was in their cognitive development—how they came to *consciousness* of the world around them—and his particular genius was in recognizing that their "errors" were as revelatory as their developmental "successes."

Piaget (and the developmental theorists who have followed him) worked from four major assumptions:

1. Children are naturally motivated to think, to learn, and to understand.
2. Children perceive and understand the world in ways that are *qualitatively* different from the ways in which adults perceive and understand the world. It is not that they know the world imperfectly or incompletely, but *differently* compared to adults.
3. Knowledge is organized into *schemas* (*schemata*). A schema is a *cognitive structure* by which things are categorized and understood. For example, we soon come to recognize the ball we see as one of a class of objects (a schema) of balls, which share certain properties. Much of learning consists in developing progressively more elaborate and sophisticated schemas (tennis balls, baseballs, footballs, etc.).
4. Learning consists of *assimilation* and *accommodation.* Assimilation is learning based on prior learning, whereas accommodation is learning created by new objects, events, and experience. Assimilation incorporates new objects and experiences into existing schemas. Accommodation requires the extension of existing schemas or the creation of new ones.

From these assumptions and based on what he observed in his own children, Piaget outlined four major stages of development:

1. *Sensorimotor stage*: Spanning birth to age two, infants develop schemas based on how the world works. They work on mastering their own innate physical reflexes, but also extending them into pleasurable or interesting actions that reach out to the external world. It is during this stage that the *egocentric* infant becomes aware of him- or herself as a separate physical entity and comes to

realize that surrounding objects have a separate and permanent existence. Piaget's entire theory of development might be described as an outline of children's journey out of egocentrism—the tendency of infants to see things from their own perspective, to see themselves as the center of the universe (indeed, at first, as *the* universe). *Object permanence*, the understanding that things have a separate and permanent existence (even when they are out of sight) is a major milestone of this stage, indicating a developmental step away from egocentrism.

2. *Preoperational stage:* During this stage, which lasts from age two to age six or seven, children learn to manipulate their environment symbolically through *inner* representations—thoughts—about the external world. It is during this stage that children begin to represent objects with words and then to manipulate those words—not the objects themselves—mentally. Major milestones of this stage include *conservation* and *reversibility*. At the beginning of the preoperational stage, a researcher shows a child two beakers, one short and fat, the other tall and thin. The child watches as water is poured from the short, fat beaker into the tall, thin one. Of course, the water rises higher in the tall, thin beaker than in the short, fat one. Even though no water has been added, the child will report that the tall, thin beaker contains more water than the short, fat one did. By the end of the preoperational stage, however, the child will recognize that the amount of water remains unchanged (the child has achieved conservation). Early in the preoperational stage, the related concept of reversibility is also absent; the child does not understand that the liquid could be poured back into the short beaker, thereby proving that the quantity is unchanged. By the end of this stage, the child understands reversibility.
3. *Concrete operations stage:* Extending from age seven to age eleven or twelve, this stage reveals the beginning of logic and of the classification of objects by their similarities and differences. Concepts of time and number also come into being during this stage. However, during this stage, children do not reason abstractly and do not employ systematic deduction or an organized approach to problems, but, rather, approach the world concretely, trying to figure things out by rote or by trial and error.
4. *Formal operations stage:* Starting at about age 12, the formal operations stage extends into adulthood. At this point, children begin to think abstractly rather than having to manipulate actual objects. Thinking becomes more orderly and systematic, allowing for *mental* experimentation. Children make and test hypotheses during this stage. Not all children—and not even all adults—attain or fully attain this stage.

Development of Moral Reasoning

Piaget's theory is cognitive; that is, it deals above all with learning. However, Piaget touched on the development of *moral reasoning*, but Lawrence Kohlberg and other psychologists carried this aspect of developmental research further. Kohlberg delineated three major stages in the development of moral reasoning: preconventional, conventional, and postconventional. The progress through these stages is from concrete to abstract; that is, children begin by basing morality on the concrete consequences of their actions. Good actions are rewarded, and bad actions punished. In the later stages of development, the sense of morality becomes progressively internalized, and children develop ethical principles by which they judge actions good or bad.

Critical and Sensitive Periods

Developmental psychologists have long been interested in *critical* or *sensitive periods*—periods during which certain skills or behaviors must be acquired or they won't be acquired later (or, perhaps, will be acquired imperfectly or only with difficulty). *Ethologists* (scientists who study the biological determinants of behavior) have studied *imprinting* in animals. In geese, for example, there is a critical period during which the gosling bonds with its mother—or whatever other moving object is present during this critical period. The gosling does not imprint before or after this period. While the situation with human beings is far more complex (nothing quite so simple as imprinting exists in humans), there is ample evidence for certain sensitive and critical periods in the areas of:

- Bonding with parents (caregivers)
- Visual perception
- Motor skills
- Language acquisition (including learning a second language)

Erikson's Life-Span Theory

Erik Erikson was a psychoanalyst, a student and colleague of Sigmund Freud, who brought Freudian theory more squarely into the social arena by developing a theory of development that emphasized social interactions and experiences. Whereas Freud continually emphasized the role of the unconscious in the development of personality, arguing that personality was essentially crystallized by middle childhood, Erikson accorded much more importance to conscious, rational processes and saw development as encompassing the entire life span. Indeed, he discerned eight distinct *ages of man*. Each of these ages is focused on a *life crisis* that must be resolved—either *adaptively* (successfully) or *maladaptively* (unsuccessfully, creating neurosis or other emotional disorder).

1. *Basic trust versus mistrust:* This is the first crisis of life during a stage that spans birth to 12 or 18 months. Infants who receive responsive caregiving develop *secure attachment* and are therefore able to achieve autonomy and relative independence. Those who receive unresponsive or abusive caregiving develop *insecure attachment*, which typically results in a fearful infant and child, who later displays little curiosity or willingness to take risks.
2. *Autonomy versus shame and doubt:* This crisis and its resolution occupy the child from about 18 months of age to 3 years old. It is during this period that the child fortunate enough to have developed secure attachment readily achieves autonomy, whereas the child who has not experienced responsive caregiving tends to experience difficulty attaining a degree of self-reliance.
3. *Initiative versus guilt:* This crisis, spanning ages 3 to 6, involves the child's need to make choices and to show initiative in attaining short-term goals. Adaptive resolution of this crisis rests on having attained autonomy in the previous stage.
4. *Industry versus inferiority:* From 6 to 12, this stage expands on the theme of initiative. Goals become longer term and more complex, involving teachers and peers. Much comparison with others occurs during this stage, providing an opportunity for the child to develop a strong, positive self-image—or a tendency to feel inferior compared with others.
5. *Identity versus role confusion:* The crisis of adolescence (from about 12 to 18 or 20) is the establishment of identity in relation to society. As Erikson saw it, adolescence is the most stressful of life's stages, even for "well-adjusted" individuals.
6. *Intimacy versus isolation:* The crisis of young adulthood—ages 20 to 40—is the resolution of marital intimacy and companionship versus the emptiness of a life of isolation.
7. *Generativity versus stagnation:* Middle adulthood (40 to 65) is characterized by generativity in the well-adapted individual—the achievement of a productive work life and the satisfactory rearing of children. Failing to achieve this generativity, life seems purposeless.
8. *Ego integrity versus despair:* The closing stage of life, from age 65 onward, is a time (at least in part) of reflection and self-assessment. Was life worthwhile? A positive answer is testament to ego integrity, while a negative answer is the equivalent of despair.

Erikson's life-span theory, his eight ages of man, is the most comprehensive theory of development yet proposed. As with other schools and major psychological theories, most modern psychologists avoid extreme partisanship. The work of

Piaget, Erikson, and Freud, as well as others, is drawn upon freely by contemporary psychologists, who use each approach as a tool to gain insight into the subject of development for the purposes of experimental study as well as clinical treatment.

CHAPTER 11

CONSCIOUSNESS, EMOTION, AND MOTIVATION

CONSCIOUSNESS IS THE COLLECTIVE ACTIVITY OF THE MIND, A CONTINUUM OF states from sleeping (and dreaming) to waking. It encompasses thought and feeling. *Emotion,* a function of consciousness, refers more specifically to feeling and to combinations of feeling and thought. *Motivation* partakes of consciousness and emotion, driving the *why* of behavior. Motivation links consciousness and, specifically, emotion to a number of basic drives essential for survival. For many psychology students—both beginning and advanced—consciousness, emotion, and motivation are the heart of the discipline: what psychology is all about.

The Levels of Consciousness

Although consciousness is a spectrum, a seamless continuum, it is useful to conceptualize it in terms of levels.

Waking

Most of us would consider this the normal level of consciousness, since it is, by definition, what we are aware of whenever we are awake. Actually, within waking consciousness, there is a range of degrees of consciousness. During a midterm or final exam, you will probably find yourself at a high level of alertness and concentration, your mental energy focused on the task before you. If you are fortunate enough to have an interesting professor, your consciousness may approach this level during lecture as well. If the gifts of the lecturer are somewhat less abundant, however, you may find yourself drifting into minimally alert levels of consciousness and daydreaming. Whereas, when you are maximally alert, your consciousness is focused on the outside world, in the daydreaming state, it turns inward, and you become aware of a *stream of consciousness*—the free flow of your own thoughts.

When you have a specific task to perform—such as listening to a dry lecture that, no matter how boring, will "be on the test" and is therefore important to you—you will doubtless find the tendency to drift into daydreaming annoying and frustrating. Nevertheless, under many conditions, daydreaming consciousness is both pleasant and creatively useful: a way of gaining access to your greater creativity.

That waking is indeed a spectrum spanning minimally alert daydreaming and a maximally alert focused state demonstrates that consciousness is a changeable balance between "awareness" of what is going on inside and outside of your body.

Subconsciousness

In everyday speech, the terms *unconscious* and *subconscious* are often used interchangeably. A psychologist, however, applies these terms to different states. The subconscious is a kind of mental border region *between* consciousness and the unconscious. At a typical level of conscious arousal, you are unaware of subconscious activity; however, this level of consciousness is readily available to your awareness if something happens to direct your attention to it.

Many "automatic" bodily functions overlap into areas of subconsciousness. Normally, for example, you are unaware of your breathing—unless you suddenly have difficulty breathing. Nor are you aware of your heartbeat—unless something frightens you, and you become aware of your pounding pulse.

Many *psychomotor activities* are also more or less automatic and subconscious. You don't have to think about moving one foot in front of the other when you walk, for example; however, if you are negotiating difficult terrain, the normally subconscious process of walking is available to your full, alert consciousness, which allows you to watch your step.

Some psychomotor activities are somewhat less firmly ensconced in the subconscious. In everyday speech, we might speak of these as "second nature." For example, tying your shoelaces: this usually requires little or no thought, unless something goes wrong with the laces (maybe one is too short). Even driving a car, a process that involves very complex acts of coordination and judgment, is to a large extent a collection of habits and well-learned behaviors. While many aspects of driving are subconscious—we don't constantly think about keeping the wheel straight; if we are driving a stick shift, we don't constantly think about just when to shift or just how to coordinate gas pedal, clutch, and shifter—we can be snapped into a level of high alertness if the situation demands it.

Selective Attention

The relationship between consciousness and subconsciousness demonstrates that our senses continually take in more than we *consciously* attend to. For example, in a crowded room, the babble of voices may recede into unintelligible background noise until you hear the mention of your name. At this, you suddenly attend *consciously* to the babble. The important point is that you have not been *consciously* listening for your name, but your subconscious nevertheless continues to process information and is, without your conscious volition, prepared to hear your name. Only when certain relevant stimuli are present does the material move from subconsciousness to consciousness. This phenomenon, called *selective attention*, and sometimes referred to as the *cocktail party phenomenon*, is very important. We move among a welter of sensory information, and we could not possibly attend to it all. We would be overloaded and unable to focus or concentrate. Selective attention tends to bring to consciousness only what is relevant.

The Unconscious

We will explore this extreme of the spectrum of consciousness in Chapter 15, when we survey the work of Freud. For it was Freud and his colleague Josef Breuer, in the late nineteenth century, who proposed the existence of this aspect of mind: a realm of mental activity "deeper" than the subconscious, and to which we normally have no access. As Freud saw it, the unconscious is the repository of painful memories, illicit desires, and other mental material too threatening to think of consciously. Yet this material is not quietly buried; it continually exerts pressure on our conscious lives, sometimes destructively so—in the form of neurosis or emotional disorder. It may also bubble up into consciousness in the form of recalled dreams (Freud considered dreams an expression of the unconscious mind and called their recollection and interpretation the "royal road to the unconscious") and so-called *Freudian slips* (also called *parapraxes*): slips of the tongue that seem to reveal what one is *unconsciously* thinking about. Freudian slips are typically of a sexual nature—and often embarrassing.

It should be observed that many recent psychologists—at least those who do not count themselves among the ranks of the psychoanalysts—minimize the importance of dreams and Freudian slips as points of access into the unconscious. And while many such psychologists acknowledge the existence of the unconscious state, they conceptualize it less as the repository of painful images and ideas than as the mental faculty that attends to basic bodily functions and other activities that do not require conscious attention.

Sleep States

Thus far we have touched only on waking states of consciousness. But most people spend about eight hours of each twenty-four—one third of the day, one-third of life—asleep. This most basic of human activities has been the subject of study for only a relatively few years, and while psychologists have learned much about the nature and function of sleep, such basic questions as why we need sleep in the first place remain unanswered. The two prominent theories hold that sleep is *recuperative*, an essentially biochemical process necessary to recovery from the waking activity of the day, or *evolutionary*—that is, a period of silent motionlessness contributed to the survival of the species, hiding our evolutionary ancestors (who, like us, could not see well in the dark) from nocturnal predators.

Rhythms and Stages

While we have no definitive answer for why we sleep, we know that sleep and waking follow regular patterns, called *circadian rhythms*, and that we suffer if these patterns are disturbed. Moreover, observation of subjects in *sleep laboratories* has revealed that sleep itself consists of regular, identifiable stages.

Circadian rhythms play a role in the regulation of metabolism and other bodily activities, such as eating, digestion, work, and play as well as sleep. When these rhythms are disturbed—as when we travel rapidly across several time zones—we experience fatigue, confusion, and even reduced resistance to disease: the symptoms of *jet lag*. Long-term deprivation of sleep can have even more dire consequences, including psychotic behavior and physical collapse.

The stages of sleep itself have been measured by the *electroencephalograph* (EEG), a device that graphically records global electrical activity of the brain.

- Sleep begins with an *initial transition to sleep*. This stage is characterized by relaxed drowsiness and, as indicated by EEG measurements, low-amplitude *alpha waves* in the range of 8 to 12 Hz (cycles per second).
- *Stage 1 sleep:* Drowsiness yields to light sleep, characterized electrically by low-amplitude *theta waves*, which have a frequency of 6 to 8 Hz. Heart rate slows, and muscles relax.
- *Stage 2 sleep:* Electrically, the amplitude of the theta waves increases, and the frequency decreases to 4 to 6 Hz. *Sleep spindles*, bursts of activity at 12 to 16 Hz, punctuate sleep during this stage.
- *Stage 3 sleep: Delta waves* supplant theta waves. Of much greater amplitude, delta waves are also much lower in frequency, ranging from 1 to 4 Hz. Viewed electrically, the impression is one of increasingly synchronized brain activity.
- *Stage 4 sleep:* This is the stage of deepest sleep, dominated by delta waves. It is difficult to wake a person from stage 4 sleep. Note that this stage is reached

early in the sleep period; that is, the sleeper normally passes rather rapidly through the first three stages, then reaches Stage 4, and spends the rest of the sleep period in progressively lighter sleep.

- *REM sleep:* From stage 4, the sleeper passes into a stage characterized by *rapid eye movement* (visible under the closed lids) and therefore labeled with the acronym *REM.* The EEG reveals brain wave patterns remarkably similar to those found in a waking and alert person—wave amplitude is lower and frequency higher—but the sleeper is, in fact, deeply asleep and, as in Stage 4, difficult to awaken. It is during REM sleep that most dreaming occurs. (Some dreams may also occur during stages 1 through 4, which are collectively labeled NREM [non-REM] sleep; however, most dreaming is associated with the REM stage.) REM sleep is also characterized by a state resembling paralysis of voluntary movement (something most of us have experienced in suddenly awaking from a disturbing dream), which may well prevent the sleeper from acting out his dream.

Dreams

As psychologists have no definitive answer to why we sleep, they have no definitive answer to why we dream. Nevertheless, it is known that:

- *All people dream.* Those who deny that they dream simply do not remember their dreams. If one of these self-professed nondreamers is awakened in the midst of a dream—say during REM sleep—he or she will, at the very least, recall having dreamed and may well recall dream content in some detail.
- *Dreaming is necessary.* If an experimental subject is deprived of REM sleep—awakened when the EEG reveals REM patterns—he or she builds up a kind of REM deficit and shows *REM rebound*: a dramatic increase in REM sleep the next night. Psychologists conclude that the person is making up for lost dream time.

Since time immemorial people have been fascinated by and have speculated on the significance of dreams. In many cultures, dreaming is regarded as a highly important and highly symbolic activity, even predictive of the future. In our own culture, we often speak of "sleeping on" an important decision or difficult problem—as if we believe that a solution will come to us in a dream. (Indeed, history is full of just such anecdotes.) Freud—and others before and after him—believed that dreams function as a kind of outlet for emotional pressures; Freud theorized specifically that these pressures were generated by the unconscious. Whereas most "enlightened," "rational" thinkers of the nineteenth century dismissed as superstition attempts to interpret the meaning of dreams symbolically, Freud theorized that the imagery of dreams was communication from the unconscious and that by interpreting these symbolic images, one could make the activity of the unconscious intelligible to consciousness. Freud spoke of the uninterpreted images of the dream as the dream's *manifest content*, which is really a symbolic disguise to mask painful *latent content*—the meaning that emerges once the symbols are interpreted properly. Much of the work of psychoanalytic therapy was and is focused on discussing and interpreting the patient's dreams. While Freud identified certain common and even universal dream symbols, symbolic types, and motifs, he believed that most images are unique to the individual and that their meaning has to be teased out through weeks, months, even years of analysis.

Sleep Disorders

Disruptions of sleep are very common, and when they become chronic, they are classified as *sleep disorders*. The disorders fall into two broad classes: *dysomnias* and *parasomnias*.

- Dysomnias include *insomnia*, difficulty falling asleep or staying asleep. *Hypersomnia* is the reverse of insomnia and is excessive sleepiness. Hypersomnia should not be confused with *narcolepsy*, in which the sufferer is episodically subject to falling asleep suddenly and unpredictably. Narcolepsy can be quite hazardous—if, for example, the narcoleptic falls asleep while driving a car. *Sleep apnea* is another common disorder, in which sleep is disturbed by difficulty breathing. The consequences of sleep apnea range from nothing more serious than loud snoring to severely disrupted sleep to a possible link between sleep apnea and *sudden infant death syndrome* (SIDS), also called *crib death*, in which infants (usually 12 months old or younger) suddenly die in their sleep.
- Parasomnias include *nightmares* (especially excessively severe or chronic nightmares), *night terrors*, and *sleepwalking*. While they may be quite disturbing, nightmares are nothing more than extremely vivid dreams, which occur during regular REM sleep. In contrast, night terrors occur during NREM sleep and are intensely terrifying (both to the sleeper and to those —such as parents—who may witness the sleeper's terror) but apparently occur without dreaming. Night terrors typically involve intense arousal, including screaming and panicked flailing of arms. Sleepwalking likewise involves apparent arousal, ranging from simply sitting up in bed and looking around, to getting up, walking, and even performing apparently purposeful activity. The belief that sleepwalkers are somehow immune from harm is groundless. Sleepwalkers are in a profound state of sleep and can indeed injure themselves. Night terrors and sleepwalking are more common in children than in adults.

Hypnotic States

No introductory psychology course is complete without a consideration of *hypnosis,* which all beginning psychology students find fascinating. Since the late eighteenth century, when it became a subject of intense interest—thanks to the work of the Austrian physician Franz Mesmer (1734–1815), from whose name the word *mesmerize* was coined—hypnosis has been subject to a great deal of controversy. It has been and is used by serious clinical practitioners to treat a wide variety of mental disorders, it has been and is used by stage magicians to entertain, and it has been and is used by out-and-out charlatans to separate the gullible from their money.

Hypnosis does produce a state of consciousness that seems different in many respects from either waking or sleeping. The *hypnotic trance,* induced by a variety of relaxation techniques, apparently produces the following behaviors in the subject:

- *Suggestibility:* Hypnotized subjects will do, think, or say things in accordance with what the hypnotist suggests. Interestingly, subjects will *not* behave in ways that are contrary to their basic beliefs. Despite what sensationalist fiction may suggest, a hypnotist cannot transform an innocent person into a criminal.
- *Dissociation from external reality:* Hypnosis apparently dissociates (separates) the subject from external reality. Hypnosis can also temporarily dissociate the subject from physical pain and from emotional distress, including the emotional distress of painful memories. See the next point.
- *Enhanced memory:* Hypnosis has been used to enable subjects to recall remote events in great detail. Police agencies have hypnotized witnesses in order to obtain details concerning crimes. In the course of psychotherapy, some patients have been hypnotized in order to bring about *age regression* in an effort to recover early childhood memories that may be relevant to therapy. Recent studies have suggested that memories "recalled" under hypnosis may contain a substantial portion of fantasy—or may even be wholly fantastic. These studies have added to the controversy surrounding hypnosis.
- *Quasi-hallucination:* Under hypnotic suggestion, the subject may be "made" to suck a lemon and imagine that it is deliciously sweet or may even be persuaded to manipulate objects that do not exist.
- *Posthypnotic suggestion:* The hypnotist may "plant" a suggestion in the subject's mind, which may linger long after the hypnotic trance ends. Hypnotherapy to help smokers kick the cigarette habit is based on this: the subject, under hypnosis, is told to associate smoking with something intensely disgusting, for example. Posthypnotic suggestions are not permanent; they fade with time, so they are only temporary solutions to therapeutic issues.

Recent studies have suggested that hypnosis does not bring about a physiologically altered state of consciousness. EEG patterns of hypnotized subjects are not significantly different from those in a state of normal consciousness. It has been suggested that hypnotic subjects in effect *imagine* that they have been hypnotized and behave in accordance with what they perceive as the hypnotist's expectations. Even if this is true, however, it is clear that hypnosis *seems real* to the subject. That is, the subject is not in any sense faking, and, therefore, hypnosis, however one views it, remains a significant manipulation of consciousness.

Drugs and Consciousness

Introductory psychology courses today typically devote significant attention to the effect of drugs, both licit and illicit, on consciousness. Drugs (including alcohol) are generally discussed in three broad categories:

1. *Depressants* slow down CNS function, reduce heart rate and respiration, impede motor function, produce sleep, and, taken in sufficient quantity, suppress bodily function to a lethal point. Depressants include alcohol, barbiturates, minor tranquilizers, and narcotics (including opiates such as opium, codeine, morphine, and heroin).
2. *Stimulants* increase CNS activity and heighten arousal and energy level, possibly bringing on euphoric feelings—or feelings of mania and panic. Taken in excessive doses or over a period of time, stimulants cause physical damage to the CNS and other organs. On an acute level, stimulant use can cause seizures or cardiac arrest. Stimulant drugs range from caffeine and nicotine to amphetamines and cocaine.
3. *Psychedelic drugs* induce hallucinations and/or perceptual distortions, heightened arousal, and euphoria. Marijuana is considered a low-grade psychedelic, whereas such substances as mescaline, psilocybin, LSD (lysergic acid diethylamide), and PCP (phencyclidene) can produce gross hallucinations and even psychotic or violent behavior.

In addition to the immediate physiological and psychological effect of drugs, the introductory psychology course will also likely consider the longer-term problems of *addiction* and *psychological dependency* as well as withdrawal. Many introductory psychology courses include material on the physical, psychological, and legal hazards associated with drug abuse.

Emotion

Most people consider the subject of emotion as central to psychology, especially since so many individuals who seek the clinical counsel of psychologists do so out of a desire to cope with painful emotions or to understand their emotions. As scientists, psychologists have developed several theories of emotion:

- The *evolutionary theory* proposed by Robert Plutchik sees the origin of emotion as adaptive in evolutionary terms; that is, emotion developed as an aid to survival. The emotion of fear, for example, motivates the individual to flee or hide from danger.

Emotions are associated with measurable changes in functions of the autonomic nervous system, including changes in heart rate, blood pressure, respiration, perspiration, and so on. Indeed, the *polygraph* (*lie detector*) does not directly detect a lie, but measures the physiological changes just mentioned, which are associated with the emotions that are, in turn, associated with lying (although not infallibly so).

While psychologists agree that emotions are linked to physiological phenomena, it is unclear just how the physiological and cognitive aspects of emotion are related and interact.

- The *James-Lange theory* holds that the physiological changes come first and then we interpret these sensations cognitively, thereby eliciting a recognizable emotion. Thus something painful or sad may make us cry and, because we are crying, we interpret our own sensations as sorrow.
- The *Cannon-Bard theory* stands James-Lange on its head, holding that situations are interpreted cognitively first, which then triggers a physiological response.
- The *Schachter-Singer theory* sees a simultaneous relationship between cognition and physiology. That is, a specific situation simultaneously elicits cognitive and physiological responses, which the individual further assesses, thereby enhancing both the cognitive and physiological dimensions of the emotion.
- The *opponent-process theory*, proposed by Richard Solomon, emphasizes a response to external stimulation that attempts to reestablish homeostasis. The stimulus is followed by the physiological and cognitive responses (an emotion or emotional *change*), which then elicits processes that tend to counteract that change in order to reestablish homeostasis. The opponent-process theory accounts for why many stimuli begin to lose their power with repetition. With repeated stimulation, the opponent processes respond more quickly and efficiently, thereby blunting the effect of the stimulus. The individual is said to develop *tolerance* to the stimulus.

Motivation

Psychologists recognize two broad categories of motivation: *primary* and *secondary.*

- Primary motives are driven by biological needs. Hunger is a primary need that will elicit behavior directed toward obtaining food.
- Secondary motives are learned, whether through direct association with primary motives or through socially mediated ways. Money, for example, is a great motivator. Money is associated with primary motives (you need money to buy food in order to eat, for example), but, through social mediation, money may also become an *autonomous motive,* a motive in its own right, as some individuals are motivated to acquire money far beyond what is necessary to satisfy their survival needs. Whether or not *aggression* is a learned motive is open to debate. Freud and others have held that it is primary, while many psychologists insist that it is learned. The motive of motivation itself—achievement for its own sake—appears to be learned, from parents, teachers, and others. Achievement motivation is often abbreviated by psychologists as *n Ach.*

Primary Motivation Disorders

Primary motives are of particular interest to clinical psychologists when they become subject to some disorder. The most common disorder of primary motivation involves eating, and the most common eating disorders involve habitual overeating, leading to obesity.

- Some psychologists believe overeating is ultimately based in evolution. Our evolutionary ancestors experienced periods of feast and famine, thus it was adaptive to evolve metabolic mechanisms for storing as much energy as possible (as fat) during periods of feast and using that energy as slowly as possible during periods of famine. In effect, human beings evolved physiological mechanisms to limit weight loss (metabolism slows during times of deprivation, and the body draws on energy stored in muscle tissue before it begins to break down fat cells), but they evolved no such mechanisms to limit weight gain.
- Overeating and obesity may also be caused by a high set point for weight. The *set point* is partially a function of the size and number of fat cells in the body and may be determined genetically or by chronic overfeeding in infancy or by some combination of the two.
- Some people convert food energy into fat more readily than others do.
- Some people respond more intensely to the sights and smells of food than others do.
- Some people eat in order to make themselves feel better emotionally.

Depending on one's background, food may be associated with comfort and love and may therefore become a substitute for comfort and love.

While overeating and obesity are the most common eating disorders, two "self-starvation" disorders are also quite common, as well as distressing, physically harmful, and potentially fatal.

- *Anorexia nervosa* is avoidance of food, resulting in sometimes extreme emaciation and death.
- *Bulimia nervosa* is an alternating cycle of food bingeing followed by purging, usually in the form of self-induced vomiting. Although bulimia does not usually lead to emaciation, chronic self-induced vomiting is unhealthy and damaging, resulting in unfavorable bodily pH balances and other side effects.

Both anorexia and bulimia are believed to be related to problems with self-image. These disorders occur most frequently in adolescent females and young women in societies (such as ours) that tend to equate attractiveness with a slender figure.

Sexual Motivation

In some ways, sex clearly qualifies as a primary motive. Reproduction, after all, is necessary to the survival of the species. However, in contrast to most other animals, human beings are not irresistibly driven to mating behavior at specific times (for example "rutting season" for deer or "heat" for cats and dogs), and human sexuality therefore partakes both of physiological drives and socially mediated motives. In human beings, sexual motivation is a compound of physical need, psychological factors, and social dictates and norms.

While it is clear that male sexual desire varies with the levels of *androgen hormones* present (though desire is not absolutely dictated by such hormone levels), there seems to be little relation between arousal and the level of *estrogen* in females. Sex, therefore, is more a matter of mutual consent than yielding to an irresistible physical force.

It is important to note that some psychologists, most prominently Freud, do indeed place more emphasis on sex as a primary motive. Freud argued that sex, although a primary motive, is, in civilized society, often *sublimated,* channeled into areas other than behavior directed toward mating or copulation. Indeed, sublimated sexuality may be seen as the driving force behind much that human beings accomplish in the arts and other creative fields.

Until fairly recently, *homosexuality* was regarded as an emotional disorder. Today, professional psychologists do not see it as a disorder—though, in many social, religious, cultural, and family situations, homosexuality is disdained and, for that reason, can create great emotional distress. While psychologists no longer treat homosexuality as a disease, they do remain interested in how homosexuality comes about. Some gay men and lesbian women report that they can trace their sexual interest in members of their sex to some childhood event, while others simply say that they have been homosexually oriented ever since they can remember. Compelling evidence exists for a genetic determinant for heterosexuality and homosexuality.

However one views the sexual motive, it is complex and complicated by many social rules, norms, prohibitions, and expectations, many of which conflict and many of which are not explicitly stated. Many people seek the counsel of psychotherapists in large part looking for answers to sexual problems.

CHAPTER 12

CONDITIONING AND LEARNING

LEARNING IS CHANGE—WE MIGHT SAY CHANGE IN THE POTENTIAL FOR BEHAVIOR—that results from experience. While *behaviorists* identify all learning as a form of *conditioning*, the majority of psychologists see it the other way around: *conditioning* is a type of learning, consisting of associations between specific stimuli and responses. We can only assume that the nature of the changes created by learning and/or conditioning are ultimately physiological in nature, though the biochemical processes involved are not yet understood.

Conditioning

The simplest form of learning—"simplest" in that the relationships involved are most directly apparent—is conditioning. There are two main types of conditioning: *classical* (also called *Pavlovian*) *conditioning* and *operant* (or *Skinnerian*) *conditioning.*

The pioneering work on conditioning was carried out by the Russian physiologist Ivan Pavlov (1949–1936), who was investigating digestion, using dogs as his laboratory models. He surgically inserted tubes into the dogs' throats to measure precisely the amount of salivation produced in response to feeding. But his attention was soon drawn away from the topic of digestion when he began to notice that the dogs often salivated at the mere approach of laboratory technicians, well *before* food was placed in their mouths. Pavlov called this phenomenon "psychic secretion," and he decided to investigate it in a systematic way.

Pavlov's earliest experiments involved purposely associating an artificial, arbitrary stimulus with a response, such that the response would be elicited any time the stimulus was introduced. He rang a bell, gave the dog food, and the dog salivated. This procedure was repeated until the dog salivated at the sound of the bell—without any food having been introduced. In the terminology of classical conditioning, the following is true:

- The food is an unconditioned (natural) stimulus (UCS).
- Salivation (with food) is an unconditioned (natural) response (UCR).
- The bell is a conditioned (artificial) stimulus (CS).
- Salivation (at the sound of the bell) is a conditioned (learned) response.

Regarding the CS, it is critically important that this stimulus be *initially neutral* with respect to the CR. That is, we know that a ringing bell does not *naturally* produce salivation in dogs.

Pavlov performed many conditioning experiments with dogs and drew many and complex conclusions. The most important include:

- Recognition of an *acquisition phase* and an *extinction phase.* In the acquisition phase, the association between the CS and CR is formed. The CS should come slightly before the UCS, so that the CS becomes a cue for the UCS. However, Pavlov found that conditioning would also occur (albeit less efficiently) if the CS and UCS were presented simultaneously. After the UCS is discontinued, the CR continues—*for a time*— then fades into *extinction.* It is *extinguished,* but may *spontaneously recover* from time to time; that is, hours or days after extinction, the CR may return upon introduction of the CS, only to disappear again.
- Generalization. Within a certain range, similar, but not identical, conditioned stimuli will all produce the same CR. A high-pitched bell will elicit salivation as readily as a low-pitched bell.
- Discrimination. It is possible to condition an animal to discriminate between two (or more) similar conditioned stimuli. If the UCS is associated exclusively with a high-pitched bell, and a low-pitched bell is exclusively associated with no UCS, the animal will respond (dog will salivate) only when the high-pitched bell is rung.
- Second- (higher-) order conditioning occurs when a previously conditioned CS is paired with a new CS. Pavlov conditioned a dog to salivate at the sound of a bell. Then he paired the sound of the bell with the lighting of a light. Soon, the light alone would elicit salivation; however, Pavlov found that this second-order conditioning was weaker than the first-order conditioning and was more quickly extinguished.

Operant Conditioning

Beginning in the late 1940s, the American psychologist B. F. Skinner reawakened interest in behaviorism by demonstrating the principle of *operant conditioning*—complex relationships between stimuli and responses, producing *operant behaviors*, by which a laboratory animal (or a person) *operates* upon its environment in order to adapt to it, control it, and get from it what is wanted or needed. Skinner saw operant conditioning as the basis of all learning—indeed, he saw it as the basis of all interaction with the environment.

- The motivation to repeat a behavior in response to some stimulus is provided by reinforcement. At its simplest, *reinforcement* is a reward that follows a certain behavior.
- Reinforcement illustrates what psychologist E. L. Thorndike called the *law of effect:* a behavior that is rewarded (leads to a satisfactory condition) tends to be repeated, whereas a behavior that is punished (leads to an unsatisfactory condition) tends not to be repeated.
- Reinforcement may be positive or negative. *Positive reinforcement* occurs when something good *(appetitive)* is presented to the subject in association with a behavior. For example, a rat presses a switch and receives food. *Negative reinforcement* occurs when something bad *(aversive)* is taken away in association with the behavior. For example, a rat presses a switch to turn off a mild electric shock. (Note that while *positive* reinforcement is a reward, *negative* reinforcement is *not* a punishment. A positive reinforcement is something added; a negative reinforcement is something taken away.
- Like reinforcement, punishment may be positive or negative. *Positive punishment* occurs when something bad (aversive) is associated with a behavior. For example, if a rat presses a switch, it is given a mild electric shock. *Negative punishment* occurs when something good (appetitive) is taken away in association with a behavior. For example, a rat is continuously supplied with food until it presses a switch.

Applied Conditioning

Skinner and his followers believed that all learning could be reduced to systematic operant conditioning. This level of complexity has not been achieved in conditioning, but the principles of conditioning are used in various learning and therapeutic situations.

Behavior modification is the application of conditioning techniques to eliminate undesirable behaviors and elicit desired behaviors. Behavior modification works well in training animals and in working with very young children or mentally retarded adults. Behavior modification techniques have also proved useful in treating *phobias*—disproportionate, uncontrollable fears of certain things or situations.

Learning Theories

While Skinner and his followers believed that all learning could be explained in terms of conditioning, other psychologists believed that learning has a significant cognitive dimension, which the behaviorist approach alone cannot adequately explain. Edward Tolman theorized that learning involves the development of *cognitive maps* without stimulus-response reinforcement. Tolman put two groups of laboratory rats through a maze. The experimental group was allowed to run freely through the maze without reinforcement. Then he removed them from the maze and added a reinforcement at the end of the maze: a goal box containing food. He timed how long it took the rats to learn the maze and obtain the food. Tolman then put the control group rats into the maze without having let them explore it in advance. Although the same reinforcement was present, it took the control group rats longer to learn the maze. They lacked the experimental group's cognitive map—which had been acquired *without* reinforcement. Thus the initial learning in the case of these rats was not motivated by stimulus and response. Tolman called this initial cognitive mapping *latent learning.* It is covert behavior, which cannot be directly measured, because there is no stimulus-and-response pattern.

Wolfgang Köhler, a prominent Gestaltist, also challenged behaviorism by demonstrating the role of *insight* in learning. In contrast to the behaviorists and to Tolman, Köhler worked with higher animals, chimpanzees. He presented a chimp with a problem, a banana suspended, out of reach, from the ceiling of the animal's cage. In the cage were a number of boxes. After attempting to reach the banana by jumping, the chimp stopped, seemed to ponder the boxes, and then stacked the boxes to reach the banana. There was no simple stimulus and response to prompt this behavior, nor was there the trial and error of Tolman's latent learning. The insight seemed sudden—and distinctly cognitive.

Synthesis

As with so much else in modern psychology, most psychologists today do not feel compelled to espouse one viewpoint or school at the expense of another. Learning, most psychologists would agree, likely involves combinations of stimulus-response behaviors, trial and error, and insight. We will survey the field and concerns of cognitive psychology further in the next chapter.

CHAPTER 13

COGNITIVE PROCESSES

IN THE MOST GENERAL SENSE, *COGNITIVE PSYCHOLOGY* CONCERNS ITSELF WITH THOSE aspects of mind that are consciously available to us, which are called *cognition*—awareness, perception, judgment, knowing, intelligence. The principal foci of cognitive psychology, however, are memory, thought, and language.

Memory

Cognitive psychologists view memory as *information processing,* offering a three-stage model that currently owes a debt to computer science:

- *Stage 1, encoding:* This is the stage of perception, in which information conveyed by the senses is converted into a form the CNS can process. As cognitive psychologists view the process, it is at this stage that *attention* comes into play: some sensory information is dropped at the sensory level, without being encoded (or perceived) at all, while other information is rejected during perception. What remains is encoded and passed on to the next stage.
- *Stage 2, storage:* As in a modern computer, the human memory, or storage system, has a short-term component (corresponding to the computer's random access memory, or RAM) and a long-term component (corresponding to the way a computer stores information on magnetic media, such as a hard disk or a diskette). Short-term memory (STM) can function as temporary storage, a kind of pass-through on the way to (more) permanent storage, or as a hold-

ing place for information that is discarded when it is no longer needed. Many memories need usefully to exist for a few moments only, as when we glance at a phone number we know we will require just once. We read it—perhaps a few digits at a time—dial it (or the group of digits), then forget it. In contrast, when we study material in preparation for an exam, we make an effort to transfer what we take in from short-term to long-term memory (LTM).

- *Stage 3, retrieval:* Cognitive psychologists are also concerned with how we retrieve (recall) stored information—as well as how we sometimes fail to retrieve it.

More About Memory Types

In addition to short-term memory and long-term memory, cognitive psychologists recognize *sensory memory,* a kind of shorter-term memory, which retains sensory information just long enough so that it can be perceived and processed further or dropped altogether. The two areas of sensory memory that have been most intensely studied are:

- *Visual sensory memory,* or *iconic memory:* Believed to function analogously to the frames of a motion picture; that is, a stream of images (icons) continually updated. Most of the "old" images are quickly dropped as they are replaced by updated images. However, many images are processed further and retained. Certain vivid images are retained in vivid detail, as when we witness some spectacular or traumatic event. These are often called *flashbulb memories.*
- *Auditory sensory memory,* or *acoustic memory:* Retains sounds (as if in brief echo) just long enough to be processed or dropped. Note that visual sensory memory is updated much more rapidly than auditory sensory memory—in terms of milliseconds versus a second or two. This is important, because sound is fleeting; a sound occurs and is gone. Most visual information, however, does not simply disappear; if a sight catches our attention, we can take another look at it and refresh our visual sensory memory long enough to decide whether the sight is worth more extended attention.

Note that the activity of visual sensory memory and auditory sensory memory takes place on a subconscious level. Short-term memory is conscious, but of limited span and capacity unless even more "conscious" effort is applied in the form of *rehearsal.* We've all found ourselves having to memorize a phone number in the absence of pencil and paper. We rehearse it, typically by repetition. Often, we rehearse material just thoroughly enough so that we can use it. The number is dialed—and then forgotten. Sometimes, however, we rehearse it more thoroughly, committing the information to long-term memory. For many people, this purposeful conversion to long-term memory requires more than simple repetition. *Elaborative rehearsal devices* include a number of *mnemonic devices,* all of which seek to associate or link the new, unfamiliar material (which you wish to memorize) with old, familiar material (which you already know). Acronyms work this way. For example, many beginning biology students find it is easier to memorize the phylogenetic hierarchy by which organisms are classified—kingdom, phylum, class, order, family, genus, species—by associating it with an acronymic phrase such as "*K*en *P*ut *C*andy *O*n *F*red's *G*reen *S*ofa." Many other associative devices exist.

Where in the brain are short-term, long-term, and sensory memory? Although physiological psychologists have identified several brain structures apparently associated with memory function, including the transfer of short-term to long-term memory, when cognitive psychologists speak of these three faculties, they are considering them in terms of their *functions,* not their anatomical *locations.*

Memories residing in long-term memory may last a lifetime, and, in contrast to the computer's hard disk, one's brain never runs out of storage space.

Long-term memories may be encoded acoustically or iconically, but the principal mode of encoding is *semantic;* that is, the encoding takes place when *meaning* is assigned to a particular memory. Many of our long-term memories are made meaningful by association with emotion or certain experiences; however, we may also assign intellectual meaning to memories. That is why it is easier to learn the facts of an academic subject in a meaningful context. For example, instead of trying to memorize facts about Freud's id concept, it is better to learn how the id functions in relation to the other two elements of Freud's concept of mind, the ego and superego. In turn, it is most effective to learn about the Freudian concept in the context of its greater significance, as a theory of mind, not just a collection of facts you need to memorize for a psychology course. Briefly stated, meaningful memories are retained more readily than memories unassociated with any particular meaning.

Forgetting

Even long-term memory is quite imperfect, and most memories are subject to decay or forgetting. Some psychologists have speculated simply that memory is rather like a *leaky bucket;* it holds information, but some of it drains out over time. Other psychologists have theorized that established memories—that is, memories that have been fully encoded into LTM—are indeed permanent, but that they cannot always be readily retrieved. This, the *interference theory,* gains support by recall under hypnosis (though some of these "memories" may be in part or in whole fictitious), recall aided by such drugs as sodium pentothal, and recall elicited by electrical stimulation of certain areas of the brain's cortex during surgery on the brain. (Patients are typically conscious during brain surgery and can report their experiences when parts of the cortex are electrically stimulated with a probe. Such work is the basis for cortical mapping, mentioned in Chapter 9.)

Interference theory seems a more adequate explanation of forgetting in the long term, and decay (the "leaky bucket") seems more adequate to explain forgetting in the short term. Memories that aren't rehearsed quickly decay in short-term memory. Memories that are rehearsed and committed to long-term memory are indeed subject to interference:

- *Proactive interference* occurs when a prior memory interferes with a newer one.
- *Retroactive interference* occurs when a new memory interferes with an older one.

One of the most common—and annoying—forms of interference is experienced as the *tip-of-the-tongue phenomenon.* You feel certain that you know something, but you can't recall it at the moment. Something—who knows what?—is temporarily interfering with retrieval. Usually, the interference will pass, and you will recall the information later—often when you are thinking about something else.

The kinds of forgetting just discussed, while frequently vexing, are normal; however, a host of diseases, disorders, and injuries can cause pathological forgetting called *amnesia.* Amnesia caused by injury, such as a blow to the head, may be *retrograde* (loss of memory after the trauma) or *anterograde* (loss of memory before the trauma). Retrograde amnesia is usually limited in temporal scope and, indeed, reversible, whereas anterograde amnesia is usually more extensive and may not be reversible. It is often a symptom of permanent damage to the brain.

Thought

Thought is, of course, a very broad and quite abstract subject. Cognitive psychologists generally deal with thought by zeroing in more concretely on issues of *problem solving.* Whereas behaviorists would argue that problems are solved, ultimately, through a chain of stimulus and response, behaviors that are reinforced or punished, the cognitive approach accords conscious processes far more of a role.

Cognitive psychologists see the basis of problem solving as an exercise of *reasoning,* either *inductively* or *deductively.*

- Inductive reasoning draws general conclusions from specific examples. It is similar to the creation and manipulation of *schemas* that developmental psychologists speak of (Chapter 10). If every ball you have seen is spherical, you might conclude that all balls are spherical. (You wouldn't be quite right, would you? Think of a football.)
- Deductive reasoning begins with a general conclusion (or hypothesis) and applies it to specific cases. If you see a spherical object, you might conclude that it is a ball.

Within these two general patterns of problem solving, we may take any number of approaches; however, most successful problem solving involves three main steps:

1. *Representation:* We first identify the main components or issues of the problem and, based on this, choose a strategy for solving the problem. The specific strategy we choose is based on our experience with problem solving in general and with solving the specific problem at hand. At this stage, the most common failure is *functional fixedness,* a failure to represent the problem fully.
2. *Generating solutions:* Next, we search our minds for possible solutions, mentally trying to fit to the problem each solution proposed. The solution may come immediately, quickly, or may require years of brain-racking mental trial and error.
3. *Evaluating the solution:* Once we recognize what appears to be an effective solution to the problem, we evaluate it, checking each feature of the solution to ensure that it addresses each aspect of the problem.

Some people are better than others at problem solving, but we all use, at one time or another, any or all of the following problem-solving strategies:

- *Trial and error:* This is both the simplest and the most arduous and time-consuming. It is essentially without plan. We simply try one solution after another until we find one that works.
- *Analysis of means and ends:* We first figure out (if we don't already know) where we are, then where we want to be. Now we fill in the steps between the two. This works well when we literally want to get from place to place, say Chicago to Los Angeles, but it can also be applied to any situation we wish to change: I am a freshman college student (that's where I'm at now). I want to be a lawyer (that's where I want to be). What do I need to do to get from where I'm at now to where I want to be?
- *Backing up:* Instead of starting with point A and looking at point B, then trying to figure out the steps between the two, it is sometimes easier to work backward from a goal. This is especially useful in devising schedules, when you have a target completion date in mind. You may work back from that date to schedule the intermediate steps.
- *Algorithms:* Many problems may be solved without having to "reinvent the wheel." We use preexisting rules to pave our way to a solution. This is the most common (and efficient) way of solving mathematical problems.
- *Heuristics:* Not all problems may be solved with handy algorithms, which are rules that *guarantee* a solution. However, the solutions to many problems can at least be guided by rules of thumb, which don't guarantee a solution, but do aid us in reaching one. You flip your bedroom light switch. Nothing happens. You conclude that the bulb has burned out—because, as a *rule of thumb,* a burned out bulb is the usual reason for a light failing to turn on. You replace

the bulb. Are you guaranteed success? No. It could be a blown fuse or a general power failure or you might have neglected to pay your electric bill, and so on.

Language

You may be surprised to find a portion of your introductory psychology course devoted to language and language development, but *psycholinguistics* has been an area of intense inquiry in psychology for many years, especially since the work of Noam Chomsky in the 1950s and 1960s. Chomsky argued that language is acquired through the operation of an innate *language acquisition device* (LAD).

You will be exposed to some basic linguistics, including the following key terms and concepts:

- *Phoneme:* The smallest unit of sound in a language. There are about 45 phonemes in English.
- *Morpheme:* The smallest unit of meaning in a language. Morphemes include whole words as well as meaningful parts of words (suffixes, prefixes, indications of tense, and so on).
- *Phrases and sentences:* More complex units of meaning consisting of combinations of words. Generally, a sentence is a complete thought.
- *Syntax:* The rules (which have some flexibility) by which sentences are assembled.
- *Semantics:* The rules (also fairly flexible) by which meaning is conveyed.
- *Grammar:* As psycholinguists understand it, the entire system of rules of a language.

Acquisition and Development of Language

The concepts just mentioned are universally used in describing the acquisition and development of language; however, psychologists tend to differ on the question of how much of language development is a function of innate processes and how much is a matter of conditioning (Chapter 12). It is a fact that all human beings experience certain observable stages of language development:

- *Newborn:* Newborn vocalizations bear little resemblance to language; however, it is clear that neonates prefer human speech to other sounds.
- *Early infancy:* Crying is the baby's primary form of communication, and parents soon learn to discriminate among crying that indicates hunger, a dirty diaper, pain, and so on. The development of communication on this level is probably largely a matter of conditioning and reinforcement.

- *Middle of first year:* Babbling, resembling human speech, appears. That babbling is universal suggests that it is an innate feature of language development.
- *6 to 10 months:* Babbling increasingly resembles speech.
- *End of first year:* Intelligible words begin to emerge from the babbling—though infants understand few words they hear.
- *Age 1 to 2: Holophrastic utterances*—single words—appear. Children may say *milk* when they want a drink. Soon, holophrastic utterances give way to *telegraphic utterances,* combinations of verbs and nouns without the connecting words or subject, as in "Want milk!"
- *Age 2 and on:* Full sentences emerge. Moreover, vocabulary increases, as does the complexity of syntax.

Syntax seems to develop from an innate sense of grammar, whereas semantics (vocabulary) is largely a matter of conditioned learning. Thus language acquisition and development is probably an example of the interaction of LAD (the innate *language acquisition device*) and conditioned learning.

CHAPTER 14

INTELLIGENCE

PSYCHOLOGISTS TYPICALLY TREAT THE AREA OF INTELLIGENCE SEPARATELY FROM cognition. For the psychologist, *intelligence* is not synonymous with cognition, but, rather, encompasses a range of cognitive *abilities.* Inasmuch as intelligence is an emotionally and culturally loaded concept, the definition and measurement of intelligence have given rise to a great deal of controversy in psychological, educational, and even political circles.

Definitions

Psychologists have developed a number of definitions of intelligence, typically in an effort to identify variables to test and measure.

David Wechsler, a pioneer in intelligence testing, defined intelligence as a general trait: the ability to act purposefully, think rationally, and deal effectively with the environment. Charles Spearman emphasized the generality of intelligence by identifying a single factor (which he labeled *g*) underlying all cognitive abilities. Although specific cognitive abilities (*s*) may vary in any individual (a person may be quite good at math, but mediocre in language, for example), that individual's *g* will generally have a bearing on all his or her *s*'s.

Some other prominent psychologists, notably L. L. Thurstone and J. P. Guilford, have argued that intelligence is a collection of essentially independent specific cognitive abilities, with no general intelligence trait underlying them. Thurstone identified seven areas—verbal comprehension, verbal fluency, numerical reasoning, abstract reasoning, spatial visualization, perceptual speed, and overall memory—all of which may vary independently and without reference to any underlying *g* factor. Guilford identified three independent areas: contents (the material we think about), operations (how we think about this material), and products (the results of our thinking). Guilford went on to subdivide these three basic dimensions according to intellectual task and came up with no fewer than 120 specific kinds of intelligence.

The *triarchic theory* of Robert Sternberg also defines three broad areas of intelligence—contextual, experiential, and componential—but finds that the three areas are interrelated to some degree. Thus the triarchic theory occupies a middle ground between those who define intelligence in terms of a general trait and those who divide it into independent specific traits.

IQ

Recall that, as a science, psychology seeks to quantify and measure what it sets out to study. It is logical, then, that psychologists would seek to measure intelligence. However, intelligence testing in the field of psychology has never been a purely scientific undertaking, and this area has been freighted with much controversy and debate.

Intelligence testing began in France early in the twentieth century with the work of Alfred Binet (1857–1911) and his collaborator Theodore Simon. In 1904, the minister of public instruction in Paris named Binet and Simon to a commission to create tests to ensure that mentally retarded children received an adequate education. (The minister was also concerned that teachers shunted off misbehaving children to classes for the retarded because they simply wanted them out of their classrooms!) Thus the first significant intelligence tests were created to serve a specific, pragmatic purpose—and this has been the impetus behind most intelligence measurement ever since.

Binet and Simon set about creating tests made up of questions they believed children of certain ages should be able to answer. Thus, the test score would yield a *mental age* (MA), which could be compared to the child's *chronological age* (CA). If a child's MA and CA pretty closely corresponded, the child's intelligence was judged to be normal. If MA significantly exceeded CA, the child was judged to be gifted. If, however, MA was significantly below CA, the child was judged retarded—and, presumably, placed in an appropriate classroom or school.

In 1916, Stanford University psychologist Lewis Terman (1877–1956) published the *Stanford-Binet Intelligence Test,* a revision of the Binet-Simon test, which yielded a more formal score than the original. Terman divided MA by CA, then multiplied the quotient by 100 to yield what he called the child's *intelligence quotient* or IQ. If a child's MA and CA precisely corresponded, he would have an IQ of 100—an "average" or "normal" IQ. An IQ significantly above or below this number indicated giftedness or retardation.

Over the years, IQ became a kind of mantra among educators as well as parents, who cooed over or cursed their children, depending on their IQ scores. As testing became more sophisticated, however, psychologists questioned whether children's intellectual growth was as orderly as Binet and Terman assumed, so that MA might not be a very meaningful concept. Moreover, how does one apply the IQ concept to adults? To address such issues, David Wechsler (1896–1981) created in 1939 the *Wechsler-Bellevue Intelligence Scale,* which was originally intended specifically to measure adult intelligence. Accordingly, Wechsler rejected the idea of an abstract or ideal mental age and instead defined normal intelligence as the actual mean test score for all members of an age group. Once derived, this mean could be represented as 100 on a standard scale, and deviations above or below it would be more meaningful than variations above or below an idealized abstraction.

Other Test Issues

Wechsler had addressed an important statistical assumption, but the quality—the content—of the tests themselves was (and remains) subject to controversy. Perhaps more than in any other field, intelligence has challenged psychologists to create tests that are *reliable, valid,* and well *standardized.*

- *Reliability* is the degree to which a test can be considered consistent. Reliability may be evaluated by testing and retesting. Subjects take the same test on two different occasions. If the test is reliable, the two scores will be approximately equal. Reliability may also be tested by the *split-half* technique, in which subjects take the test once, but odd-numbered items are scored separately from even-numbered items, and the two scores are compared. They should be approximately equal.
- *Validity* is more difficult to assess. The question of validity is simple enough: does the test truly measure what it sets out to measure? However, answering the question is not so simple. Evaluating validity requires an assessment of *face value* (on the surface, do the questions seem appropriate?), *content* (are the questions sufficiently representative of what constitutes intelligence?), *criterion validity* (do the subjects' scores on the test in question correspond to their scores on other intelligence tests?). Finally, there is the issue of *construct validity,* which asks what, precisely, is actually being measured. Where intelligence is concerned, this remains subject to debate.
- *Standardization* comprises two considerations. First, test *procedures* must be standardized: the test must be administered the same way in all cases. Second, the individual's score must be compared to *norms,* which are established as a result of having administered the tests to many large groups.

Special Cases

So far we have surveyed the general issue of intelligence and how it is measured. Intelligence spanning a broad range is considered normal, but individuals who score at the extreme low end are considered *mentally retarded,* while those who score at the extreme high end are considered *gifted.*

Retardation

A diagnosis of mental retardation is based on three criteria:

1. Intellectual function is significantly below average. On the Wechsler tests, retardation begins at an IQ of 70 and below. Using this criterion, approximately 2 percent of the population is mentally retarded.
2. Adaptive behavior is significantly impaired. Skills essential to taking care of oneself and functioning from day to day have been inventoried on standardized checklists, which teachers and other caregivers can fill out.
3. The onset of retardation must come before age 18, which is conventionally considered the end of the developmental period. Impairment of intellectual functioning that occurs after age 18 must be attributed to a cause other than mental retardation—perhaps disease or injury.

Once a diagnosis of mental retardation is made, caregivers and educators find it useful to assess the degree of retardation. IQ is the principal measure of degree, supplemented, however, by assessment of adaptive behavior. A Wechsler score between 55 and 70 indicates mild retardation; 40 to 55 is considered moderate retardation; 25 to 40 is severe retardation; and those with scores below 25 are considered profoundly retarded. People with mild retardation are generally *educable* and can live with a high degree of independence. Moderately retarded people are *trainable* and typically require some form of assistance in daily life. The severely and profoundly retarded are capable of few adaptive behaviors and generally require the constant supervision of custodial care.

The causes of mental retardation fall into two broad categories: *organic* and *psychosocial.* By far the more common causes are organic and include genetic abnormalities; birth defects caused by teratogens, illness, or injury; brain damage that occurs during birth; and neurological disease or trauma in infancy or childhood. Psychosocial causes of retardation include various forms of severe intellectual and social deprivation and abuse.

Learning Disorders

Psychologists and educators are careful to distinguish *learning disorders* from mental retardation. Individuals with learning disorders have specific areas of deficiency, but function normally in overall cognitive ability.

Giftedness

As mental retardation characterizes about 2 percent of the population, so does the opposite extreme, *giftedness,* which is defined as a Wechsler IQ of 130 or greater. In practice, admission to gifted programs in school is based on classroom performance and teacher recommendation in addition to IQ test results. It is certainly possible to have a high IQ and lack the motivation to prosper in a gifted program.

Can IQ Be Enhanced?

Again, psychologists differ in their opinions about the degree to which IQ can be enhanced versus the degree to which it is innate and, therefore, relatively fixed. It is important to distinguish between intelligence and IQ. Indeed, there is no question that intelligence increases into adulthood as we learn more. IQ, however, is a measure of intelligence *relative to age,* so is designed to remain reasonably constant throughout life. This notwithstanding, a number of studies have found that IQ can change.

Genetics Versus Environment

As is so often the case with nature-versus-nurture arguments, the truth seems to lie somewhere between the extreme positions. IQ of children does correlate strongly with the IQ of the parents, even in cases where twins have been adopted and reared apart. In such cases, the IQ of the twins correlates more closely with that of the natural parents than with that of the adoptive parents. Nevertheless, environmental factors do influence the level of IQ:

- Good health care and nutrition are critical.
- A stimulating environment, especially in infancy and early childhood, is important.
- Positive social interaction is important, especially *responsive caregiving.*
- Preschool programs (such as Project Head Start) enhance IQ.

Note that these environmental influences don't necessarily resolve the nature-versus-nurture debate. It may be argued that a maximum potential IQ is inherited as a genotype and that environmental factors largely determine the degree to which that potential is manifested as a phenotype (see Chapter 9), but that environment will not increase IQ beyond what has been genetically ordained.

CHAPTER 15

PERSONALITY

PERSONALITY IS THE PRIMARY FOCUS OF SIGMUND FREUD'S THEORY OF PSYCHO-analysis, and no theorist has yet created a more comprehensive theory of personality than Freud. However, many psychologists have since revised and departed from Freudian theory and have also challenged it in profound ways. Some advocates of social psychology have even attacked the idea of personality itself, suggesting that it is not a property of individual psychology at all. In fact, some social psychology theorists suggest that behavior is so strongly influenced by a given social context or situation that personality is nothing more than an illusion or even a comforting fiction, the product of a need to feel a sense of continuity of self.

Psychoanalysis

Sigmund Freud developed his theory of personality through a combination of clinical work with his patients and introspection into himself, both processes leavened and flavored by his deep interest in classical literature, mythology, and religion (all of which Freud regarded as sources of insight into human behavior). Freud outlined two *developmental progressions* in the creation of personality. The first of these reveals the basic *tripartite structure* of personality, as Freud conceived it. The second of these progressions is a journey through *psychosexual stages* spanning infancy to puberty.

The Tripartite Structure of Personality

To begin with, we should observe that *structure* is used here for lack of a better word.

Freud's *id, ego,* and *superego* are not physical or physiological structures, but dynamic systems, which have no particular seat in any identifiable part of the brain.

The first of these systems, the id, is innate, present at birth, the source of *instinctual wishes* driven by biological needs. Freud saw two categories of instinctual wishes: *eros* (the life instincts) and *thanatos* (the death instincts). Thus, in Freud's model of personality, we are born with opposing forces at the root of ourselves: instincts associated with satisfying life needs (hunger, thirst, sex, and so on) and instincts associated with self-destruction (including masochism and aggression). The imperatives of the id are strong. It seeks immediate gratification of its wishes (the *pleasure principle*).

The id is with us for life. It is part of the *unconscious mind* and, therefore, ordinarily unavailable to cognition—yet always pressing us to satisfy its demands.

As infants grow into childhood, they develop another personality system, the ego. Whereas the id operates according to the pleasure principle, the *ego*—essentially synonymous with consciousness, the conscious concept of the self—operates according to the *reality principle.* That is, the ego straddles the id and the world. It is driven by the pressure of the id to find ways to gratify the wishes of the id; however, in accordance with the reality principle, the ego recognizes that it cannot always gratify these wishes. When such gratification is not possible, the ego calls upon *defense mechanisms.*

Freud (and some of his disciples) identified a variety of defense mechanisms, of which *repression* is the most basic. Those id-borne wishes the ego finds unacceptable (painful, offensive, and so on), it represses, relegating them to the unconscious. This repression is itself unconscious—automatic—and occurs without any awareness or volition. Although repressed and unavailable to the ego, these wishes do not disappear. They continue to exert pressure, sometimes in emotionally and even physically destructive ways. They may express themselves in symbolically disguised form through such vehicles as dream images or slips of the tongue.

In addition to repression, other defense mechanisms are also important:

- *Rationalization:* We are adept at talking ourselves into believing we *want to do* something when we really don't want to, or we *don't want to do* something when we really do. For example, the man who fails to get the lucrative job he applied for may rationalize that he wouldn't have been happy in that job anyway.
- *Denial:* We sometimes simply refuse to accept an unpleasant reality. The wanna-be singer who is repeatedly told she has no musical talent may ignore such criticism and lavish hard-earned money on more and more voice lessons.
- *Projection:* We are often skilled at blaming others for our own faults. A man who eats chocolates to the point of obesity does not blame himself but curses "those darn candy makers."

- *Displacement:* We sometimes direct our unpleasant feelings at alternate targets. A man's boss enrages him, but if he lashes out at the boss, he may lose his job. Instead, then, he goes home and hollers at his wife and children.
- *Reaction formation:* In Shakespeare's *Macbeth,* Lady Macbeth's guilt is suspected when the "lady doth protest too much"—acts too innocent. We sometimes behave or speak in a manner opposite our true feelings —when our true feelings are unacceptable to us. Thus the man with a taste for pornography becomes the town's most ardent crusader on behalf of "moral decency."

Most defense mechanisms are fundamentally duplicitous and, therefore, stressful, exacting on us an emotional cost that may even result in mental disorder (neurosis). However, a few defense mechanisms are reasonably positive or even adaptive.

- *Sublimation* is socially acceptable displacement of, for example, sexual desire or aggression. Freud believed that many of civilization's greatest accomplishments—in the arts, in literature, in music, and so on—are driven by sublimated sexual desire. And anyone who watches Monday-night football may readily conclude that the sport is driven by sublimated aggression.
- *Intellectualization* is a process of emotionally detaching oneself from an emotionally wrenching situation. There are times when this behavior is destructive—as when a parent is habitually remote and unemotional in relations with his or her child. But there are times when such detachment is important, as when a judge must make an impartial decision on a question of law or when a paramedic must deal efficiently and rationally with the victims of a terrible automobile accident.

The third element of the tripartite personality develops after infancy but early in childhood, before the child reaches school age. The *superego* roughly corresponds to conscience as an internalized sense of ethics or right and wrong. Whereas the id operates according to the pleasure principle and the ego according to the reality principle, the superego operates according to the *morality principle.* As one matures, it is the superego that most often comes into conflict with the wishes of the id. If the superego cannot coax the ego into acting in accordance with its morality principle and against the pleasure principle of the id, the result is *guilt.*

For Freud, emotional health and personal fulfillment are largely matters of striking a balance among the three aspects of personality, id, ego, and superego—no easy task, since the relation among these elements is never static, but always in flux.

The Psychosexual Stages

Allied with the tripartite structure of personality is a process of progression through the *psychosexual stages.* This progression is more concrete, more dramatically a relation of mind and body, than the progression through the three levels of personality.

Each psychosexual stage is a shift from one *erogenous zone* to another. Recall that the id operates according to the pleasure principle. In infancy, this means that the id drives the baby to derive pleasure from his or her body. In earliest infancy, the focus of such gratification is oral—the lips and oral cavity constitute the infant's erogenous zone, and, therefore, the first psychosexual stage is the *oral stage*. Note that an oral erogenous zone is highly adaptive in an infant, whose nourishment comes from the mother's breast, which the infant seeks with his or her mouth. Without this drive to seek and suck at the breast, the infant might well fail to survive.

The oral stage does not simply disappear with maturity. How the infant's desire for oral pleasure is gratified has a lasting effect on personality. If overfed or underfed, for example, the individual may become *orally fixated*. This might manifest itself in smoking, in overeating, or perhaps in a compulsive desire to talk.

As infants progress to toddlerhood, they pass from the oral to the *anal stage*, during which they derive pleasure from activity associated with elimination of feces. According to Freud, toilet training, which typically takes place during this period, can have profound effects on adult personality. Strict, punitive toilet training may produce an *anal-retentive personality*, characterized in the adult by excessive fastidiousness, obsession with cleanliness, mean-spiritedness, difficulty expressing emotion, exaggerated attention to detail at the expense of the big picture or true creativity. On the other hand, lax toilet training may produce an *anal-expulsive personality*, characterized by general looseness of emotions, a lack of control, flightiness, and, perhaps, uncontrolled aggressiveness.

From about the age of three onward, the erogenous zone shifts to the genitalia. Although it does so in boys as well as girls, Freud referred to this stage as the *phallic stage*. The child derives his or her chief physical pleasure from stimulating the genitals. Again, note that this is ultimately adaptive behavior, linked, at least at some evolutionary level, to reproduction.

It is during the phallic stage that boys experience the *Oedipus conflict* and girls the *Electra conflict*. Freud argued that the onset of genital sexuality leads the child to desire the parent of the opposite sex. In the case of the boy child, this desire is perceived as illicit, and the child fears that he will be found out by his father, who will castrate him. (The child does not consciously fear these things, but entertains unconscious and symbolic fantasies of such a scenario.) *Castration anxiety* ultimately leads the boy to identify with and emulate his father, partially to acquire his power, but also in the (again, unconscious) belief that identifying with his father will save him from castration. A by-product of this effort to identify with the father is the formation of the superego.

The situation with the girl child is rather different. She desires her father, and in particular covets his penis because she does not have one (*penis envy*). Ultimately, she identifies with her mother as a way of symbolically possessing the father and his penis. This identification is passive, in contrast to the fear-driven identification of the boy with the father. After all, the girl does not fear castration. Lacking the boy's strong motive for identification, the girl forms a weaker superego and, according to Freud, is therefore less moral than the male child. Freud, of course, was a product of his male-dominated age, and this aspect of psychoanalytic theory strongly betrays the bias of its cultural origin. The psychodynamicist Karen Horney (one of the Neo-Freudians) addressed many of what she felt were Freud's theoretical misconceptions about women, including castration anxiety, penis envy, and the notion that women are, perforce, morally inferior to men.

The Neo-Freudians

Psychoanalysis is one of the most compelling, pervasive, and influential ideas in modern Western thought. Many psychologists, particularly clinicians, still cleave closely to classical Freudian theory. However, following Freud, a number of psychologists modified psychoanalysis in various ways, typically directing more attention to conscious (ego) personality processes. Collectively, these Freudian-inspired revisions are called *psychodynamic theories.*

The work of one of the most important of the neo-Freudians, Erik Erikson, was addressed in Chapter 10.

Jung

Carl Gustav Jung (1875–1961) was a student of Freud who broke with the master when he expanded Freud's concept of the personal unconscious into a *collective unconscious* universal among humankind. The engagement of the individual with the universal images (called archetypes) of the collective unconscious produces personality.

Archetypes are aspects of the genetic inheritance of the species. They are forms, innate within our minds, which we expect to encounter and, therefore, spend our lives seeking. Jung identified the rich imagery and characters of mythology, primitive art, and classical literature as expressions of archetypes (the Leader, the Mother, the Savior, and so on), and he also saw each of us as fashioning similar archetypes and perceiving ourselves and the world through them.

On a personal level, Jung contrasted *self*—who we really are—with *persona* (mask), the image of self we present to others. Significant discrepancy between self and persona is a source of mental disorder. Jung also identified an equivalent of the Freudian id, which he called the *shadow,* in which such primitive or animal impulses as aggression and selfishness reside. Failure to make contact with the shadow may mean that it will exercise undue influence over conscious life. Jung also spoke of a male and a female principle (*animus* and *anima*) residing in all of us, regardless of gender. Thus, a degree of androgyny is healthy in the well-adjusted personality. As each of us contains both animus and anima, so we are also endowed with the traits of *extroversion* and *introversion,* a tendency to be outgoing and gregarious and a countervailing tendency to be inward, shy, reflective, and timid. As the healthy personality must achieve a balance between animus and anima, so it must reconcile its introverted and extroverted elements.

Adler

Another student of Freud, Alfred Adler (1870–1937) emphasized a *striving for superiority* as the prime mover behind the development of personality. The condition of the newborn and young child, Adler argued, is by definition inferior, since the infant and child have limited abilities. Thus we enter the world with *feelings of inferiority,* which we strive to overcome through *compensation*—a process of making the best of our strong points in order to overcome our weaknesses. Success leads to a balanced, highly motivated, mature personality. Failure may result in *overcompensation* (a fanatical and always frustrated desire to excel at all costs) and the development of an *inferiority complex,* a chronic sense of inferiority.

Humanistic Theories of Personality

Whereas the Neo-Freudians revised Freud, the humanist psychologists reacted distinctly against the psychoanalytic emphasis on the primitive, animalistic, and unconscious aspects of human nature, just as they reacted against the behaviorists' emphasis on automatic, will-less patterns of stimulus and response. They felt that, in different ways, both psychoanalysis and behaviorism were dehumanizing, and they opposed to these approaches their own *phenomenological approach,* which put the theoretical emphasis not on the psychologist's theory, but on the individual. The psychologist's task became understanding the individual on his or her own terms.

The two most influential humanists were Carl Rogers (1902–1987) and Abraham Maslow (1908–1970). Rogers's *person-centered theory* and its clinical counterpart, client-centered therapy, was aimed at resolving the conflict he saw as essential in personality development: the gap between the *real self* and the *ideal self.* Rogers saw personality development as a process of *actualization,* a desire to become all that we can. However, this desire becomes distorted by *conditions of worth*—unrealistic expectations imposed upon children by parents and other adults. Conditions of worth create an ideal self that cannot be realistically attained, with the result that the gap between real self and ideal self seems hopeless, creating despair as well as chronic feelings of failure and inferiority.

To damaging conditions of worth, Rogers (and other humanists) opposed *unconditional positive regard*—universal, unconditional acceptance of the individual as good. True, a particular person's behavior may be wholly unacceptable, but the therapist must separate the bad behavior from the good person underneath. Similarly, parents must give their children *unconditional love*, love that is clearly not contingent on the child's meeting any parentally established conditions of worth.

Like Rogers, Maslow accepted the values of humanist psychology, but he worked out a more comprehensive theory of personality based on a *hierarchy of needs.* The most basic needs are physiological and relate to immediate survival. Satisfaction of thirst and hunger are physiological needs. Closely related to these most basic needs are those of the next tier, *safety needs.* Shelter and avoidance of physical threats are principal safety needs. From here, the hierarchy rises to the social realm and *belongingness needs,* which relate to human social interaction and love. Higher still are *esteem needs,* the need for a positive self-image, achievement, and the respect of others. At the top of the hierarchy of needs is *self-actualization,* the development of personality at its highest, including the realization of one's own talent and self-fulfillment. The self-actualized person achieves a high quality of life.

Personality as Traits

Some relatively recent theorists have turned from defining personality in terms of drives and needs and have embraced the less dynamic concept of the *trait*—a characteristic or attribute that is variable in degree. Gordon Allport took a distinctly empirical approach to defining and classifying the traits that make up the building blocks of personality. He reviewed the English words used to describe personality traits, winnowing them to some 4500. He then sorted and classified these terms into five categories:

1. *Common traits:* Those that apply to people in general.
2. *Personal traits:* Those unique to the individual.
3. *Cardinal traits:* Traits that are manifest in an individual's behavior across all situations.
4. *Central traits:* Traits that are broadly manifest, but not in all situations.
5. *Secondary traits:* Traits manifest in an individual in certain situations only.

Thus Allport attempted to reconcile the concept of a stable personality with the social psychologist's view of what we call personality as a function of the individual's behavior in a given situation. Some traits are common to people in general and some characterize particular individuals, yet they may be manifest always, in most situations, or in some situations only.

Raymond Cattell took Allport's 4500 personality traits and winnowed it further until he had identified 16 *source traits,* which are roughly equivalent to Allport's central and cardinal traits and underlie *surface traits*—the actual overt behaviors an individual manifests as products of his or her personality.

Other, more recent psychologists have proposed additional personality trait inventories, including the so-called *Big Five* approach, which casts personality into five core dimensions: extroversion, emotional stability, agreeableness, openness, and conscientiousness.

CHAPTER 16

MENTAL ILLNESS

ASK THE AVERAGE PERSON WHAT A PSYCHOLOGIST DOES, AND YOU WILL PROBABLY BE told that a psychologist works with the mentally ill. Most people think of psychology as *clinical* psychology. While it is true that many psychologists do work with "clients" or "patients," the field of mental disorder is the province of theorists and experimentalists as well as clinicians. Collectively, the area of psychology dealing with mental illness is called *abnormal psychology.*

What's Abnormal?

Among the first tasks of abnormal psychology is to define abnormality. Often, abnormal behavior is overt and obvious. Some people look and act crazy. But the range of normality is not always rigidly defined, and it is certainly anchored to what a particular society or culture finds acceptable and unacceptable in addition to how the individual feels and acts. So, to begin with, most definitions of abnormality are made in the context of social and cultural *norms:* the generally unwritten rules of behavior in a particular society.

Beyond this, most psychologists agree that for behavior to be classified as abnormal, it must be:

- Significantly unusual or bizarre
- Significantly maladaptive
- Significantly troublesome to the self or to others

"Significantly" may be defined as frequent and severe.

What Is Unusual or Bizarre?

Significant violation of social norms is often sufficient to warrant identifying behavior as abnormal. Often, such behavior transgresses not only implicit social norms, but explicit laws, such as prohibitions against public nudity, sexual misconduct, acts of vandalism, or even acts causing injury or death. Other acts may not be illegal, but are nevertheless bizarre: gesturing wildly in public, talking to oneself in an alien tongue, howling, and so on. Some bizarre or unusual behavior, however, is *covert* rather than *overt.* Hallucinations and delusions, hearing voices, "seeing things," believing everyone is conspiring against you, believing you have a mysterious ailment, believing you are being guided or coerced by alien forces, fearing things that are normally harmless, experiencing significantly exaggerated moods, mood swings, or significantly inappropriate emotional responses—all of these are covert abnormal behaviors.

What Is Maladaptive?

Many clearly unusual or bizarre behaviors are also clearly maladaptive. The man who howls and flaps his arms like a chicken is not likely to find gainful employment. The woman who compulsively shoplifts may end up in jail. But not all maladaptive behavior is so strange or obviously "wrong." Chronic inability to function in everyday life—at home or at work—chronic anxiety, mood swings, depression, chronic difficulty even thinking—are common examples of more subtle maladaptive behavior. The severity of such behavior ranges from the *neurotic* (relatively mild, but still *in some degree* debilitating) to the *psychotic* (severely debilitating).

What Is Troublesome?

Behaviors and feelings become chronically troublesome to oneself if they significantly compromise the quality of life. The person who always feels sad or "blue" or depressed, the person who continually feels fearful, the person who is racked by unexplained anger or guilt may well seek psychological counseling. In severe cases, such troublesome behavior may lead to suicide. A small minority of severely disturbed people pose a danger not just to themselves but to others and are capable of causing injury or even death.

Types of Disorders

To the nonprofessional, mental illness may present a confusing and frightening picture. An important step toward understanding—and, ultimately, treating—mental illness is to describe and classify the various forms it takes. This has long been an area of some controversy, and, indeed, some psychiatrists and psychologists reject the term *mental illness* in many instances and prefer *abnormal behavior* or *behavior disorder* instead. However, the most widely accepted catalog and classification of mental illnesses is the American Psychiatric Association's *Diagnostic and Statistical Manual of Mental Disorders* (4th edition), usually abbreviated DSM-IV.

DSM-IV classifies mental disorders into the following broad categories:

A *psychiatrist* is a medical doctor (M.D.) who specializes in the diagnosis and treatment of mental illness. A *psychologist* is not a medical doctor, but often holds a Ph.D. degree. A *clinical psychologist* specializes in the diagnosis and treatment of mental illness. In contrast to a psychiatrist, a psychologist is not permitted by law to prescribe medication.

- *Disorders diagnosed in infancy, childhood, and adolescence:* These include mental retardation, autism, and learning disorders.
- *Delirium and dementia:* Disorders in this group are often associated with specific diseases (such as Alzheimer's).
- *Disorders related to substance abuse:* Addiction and the various adverse effects of psychoactive drugs as well as certain toxic substances are included in this category.
- *Schizophrenia and other psychotic disorders:* These are the severely abnormal and maladaptive mental illnesses, of which schizophrenia is the most varied and important.
- *Mood disorders:* Major depression and bipolar disorder (formerly called manic-depression) are the chief mood disorders.
- *Anxiety disorders:* These include chronic feelings of generalized anxiety, panic attacks, and phobic conditions.
- *Somatoform disorders:* The most common disorders in this classification are hypochondriasis and various psychosomatic complaints. These are not imaginary illnesses, but very real illnesses caused by emotional problems rather than organic problems.
- *Dissociative disorders:* These problems range from psychogenic amnesia to "multiple personality."
- *Sexual and gender identity disorders:* These include sexual dysfunction of psychogenic (rather than organic) origin, and sexual deviation that significantly interferes with a person's life. Note that DSM-IV does not classify homosexuality as a disorder.
- *Eating disorders:* The class of disorders that includes anorexia nervosa and bulimia nervosa.
- *Sleep disorders:* Includes sleep problems that significantly interfere with a person's life, ranging from chronic insomnia to narcolepsy.
- *Impulse-control disorders:* Various compulsions are classed in this category, including kleptomania, pyromania, and pathological gambling.
- *Adjustment disorders:* Covers a wide range of symptoms caused by failure to adjust to stressors (events and situations that affect us psychologically). Post-

traumatic disorder (lingering emotional disturbance after some traumatic event) is included in this category.

- *Personality disorders:* Maladaptive and troublesome personality traits are included in this category.

Major Categories: A Closer Look

Abnormal psychology is a vast and complex field, which introductory courses can do no more than survey. You are likely to take a closer look at five categories of disorder: mood disorders, anxiety disorders, dissociative disorders, schizophrenia, and personality disorders. These are the most common and broadest areas.

Mood Disorders

Mood, as used in clinical psychology, refers to a person's overall emotional state. Contrast the term *affect,* which refers to transient emotional states, such as giggling at a joke or feeling sadness when some painful event occurs. In *mood disorders,* exaggerated states of emotionality disrupt a person's life and functioning.

- Depression or *major depressive disorder* is a state of chronic, deep negativity, characterized by unremitting sadness, feelings of guilt and worthlessness, and a general feeling of hopelessness. The condition may disable a person from normal daily functioning and may provoke suicide.
- *Bipolar disorder* used to be called manic-depression. This common mental illness consists of cyclic mood swings from manic states (ranging from high energy and hyperactivity to increased irritability) to profound depression. In some individuals, the depressive portion of the cycle predominates, in others it is the manic, and in others still the two extremes figure fairly equally.

Anxiety Disorders

Everyone feels anxious from time to time, but chronic anxiety without *apparent* external cause is considered abnormal and often prompts sufferers to seek professional counsel.

- *Generalized anxiety disorder* is marked by chronic feelings of anxiety possibly punctuated by sudden and overwhelming *panic attacks.*
- *Phobias* are overwhelming, unreasoning fears typically associated with specific situations or objects. Phobias can be crippling—as when an airline pilot develops a phobia related to flying or when anyone develops *agoraphobia,* a fear of open spaces or crowds, which may compel the sufferer to remain locked within his or her home. Depending on the individual, phobias may be

associated with virtually any situation or object one may think of, but some of the most common are agoraphobia (just mentioned) and acrophobia (fear of heights), plus phobias common in modern life, such as fear of riding on elevators or fear of flying.

Dissociative Disorders

Dissociation is the separation of functions normally related. Dissociations of functions can take place within the cognitive realm or within the realm of personality.

The type of dissociation related primarily to cognition is *dissociative amnesia,* severe memory loss due to psychogenic causes (that is, unlike anterograde and retrograde amnesia, discussed in Chapter 9, *not* caused by physical trauma).

Related primarily to the realm of personality are *depersonalization disorder* and *dissociative identity disorder* (also called *multiple personality disorder*).

- Depersonalization disorder is a chronic and disturbing feeling of separation of mind and body, as if one were outside of oneself, looking in, without control over thoughts and actions.
- Dissociative identity disorder involves the "splitting" of the self into two or more personalities —typically personalities with different names and very different traits and feelings. Popularly called "split personality," this disorder is also sometimes confused with schizophrenia (also popularly miscalled "split personality"). The two disorders are unrelated.

Schizophrenia

Popularly misnamed "split personality," schizophrenia does come from a Greek compound word signifying split mental functions. This is not to be confused with the dissociative disorders, however, because the split involved is far more profound. The schizophrenic suffers a split between perception and thought. The symptoms of the disorder may be continuous and unremitting or may be episodic. In severe cases, schizophrenia is characterized by tragically disabling disruptions in the most basic cognitive processes. Typical symptoms include:

- "*Word salads*"*:* a jumbled, quasi-nonsensical flow of words, some of which may be recognizable items of vocabulary and some of which may be *neologisms,* made-up words. The schizophrenic may also *perseverate,* meaninglessly repeating words; *echolalia* refers to the meaningless repetition of the words of others.
- *Hallucinations:* The schizophrenic may see and hear things that do not physically exist. Auditory hallucinations are more common than visual hallucina-

tions; schizophrenics often report hearing voices, which may command them to do or say various things or to feel a certain way.

- *Delusions:* In everyday speech, the terms *hallucination* and *delusion* are often confused. *Hallucination* is defined above; *delusion* refers not to perceiving sights or sounds, but to holding beliefs that have no basis in reality. Schizophrenics often suffer *delusions of influence,* believing that someone or some external presence is controlling their thoughts. They may also suffer *delusions of persecution,* the belief that others are conspiring against them. This is the case in *paranoid schizophrenia,* the most common form of schizophrenia. Some schizophrenics harbor *delusions of grandeur,* a belief that they are supremely important, the incarnation of a god (or God), or even some famous historical personage (such as Napoleon).

In addition to these most dramatic symptoms, schizophrenics suffer from *disturbances of affect.* That is, they exhibit *flattened affect* (emotionlessness) or *inappropriate affect* (such as laughing at a funeral). Schizophrenics are also typically withdrawn socially—in large part because their condition isolates them.

While most schizophrenics suffer severe disruption of the ability to function on an everyday basis, *catatonic schizophrenics* are wholly disabled. Their condition is characterized by profound disruption of motor activity. They may sit and stare with complete immobility (true *catatonia*) or may exhibit *catalepsy,* assuming whatever pose someone else places them in, as if they were molded of wax or clay. Some catatonic schizophrenics endlessly repeat gestures and other apparently meaningless activity. Whatever the specific form catatonic schizophrenia may take, the patient is severely or completely lacking in responsiveness to others.

Personality Disorders

People with personality disorders often seem normal—except for the single trait or cluster of traits that marks their personality and that is maladaptive and troublesome to the self or to others. Some personality disorders are clearly dangerous to the self or to others. The two most common groups of personality disorders are *narcissistic personality disorder* and *antisocial personality disorder.*

- *Narcissistic personality disorder:* In Greek mythology, Narcissus was the demigod so obsessed with his own image that he gazed at himself in a pool of water until he fell in and drowned. People afflicted with narcissistic personality disorder have a grotesquely inflated attitude about their own superiority and uniqueness. Now, many "normal" people are conceited and "full of themselves," but those who display this disorder go far beyond mere conceit. They are unable to empathize with others or to respect the rights or feelings of others, and they are willing to sacrifice the well-being of others in order to gain attention and admiration.

- *Antisocial personality disorder:* While narcissists can make themselves as well as others miserable, those afflicted with antisocial personality disorder can be truly dangerous. They are wholly self-centered and freely manipulate (lie to, cheat, and steal from) others to satisfy themselves. They do so without any guilt, remorse, or empathy. In some extreme cases (called *psychopathy* or *sociopathy*), individuals with this disorder have no scruples about physically harming or murdering others if they perceive that doing so will benefit themselves.

Treatment of Mental Illness

The treatment of mental illness is a complex subject with a long and checkered (as well as quite fascinating) history. The typical introductory psychology course generally presents a brief overview of treatment approaches.

Historical Perspectives

For most of recorded history, the mentally ill were deemed to be possessed by demons, evil spirits, or the devil. While in most societies they were not regarded as criminals (indeed, it was often recognized that they were not responsible for their actions), they were typically treated as criminals—or worse. The chief "treatment" consisted of confinement to *asylums*, which were characterized by neglect (at best) or outright abuse (more typically). During the eighteenth century, Philippe Pinel (1745–1826), a Parisian physician, set about reforming and humanizing the treatment of the mentally ill. Gradually, Pinel's ideas took hold, and asylum conditions improved.

Nevertheless, even though treatment of the institutionalized mentally ill improved, they were, in effect, *warehoused*—looked after and kept from harm (or from harming others), but not intensively treated. In large part, this was because there was little that could be done to treat major mental illnesses effectively. This situation changed beginning in the 1950s and early 1960s with the introduction of powerful *psychotropic* drugs, which will be discussed momentarily. These medications prompted the *deinstitutionalization* of many severely mentally ill persons, who were subsequently treated on an outpatient basis.

Historians of mental illness devote most attention to the treatment of the severely ill. As to others, who suffered from anxiety disorders, depression, and the like—miserable, perhaps, but still able to function in society—treatment, historically, has been scarce. Physicians prescribed various tonics and other remedies, ranging from bed rest, to recreational travel, to cold baths, and the clergy recommended prayer. It wasn't until the nineteenth century and the work of Sigmund Freud that the "walking wounded" of mental illness began to receive very specific, concentrated attention.

Approaches to Treatment

Modern treatment of mental illness falls into three broad categories, which are sometimes used in combination:

- *Medical and psychiatric treatment* emphasizes the use of psychotropic drugs and, in some cases, other forms of physiological intervention.
- *Psychotherapeutic* approaches emphasize analysis of problems and counseling.
- *Behaviorist* approaches use conditioning techniques to modify undesirable behavior without reference to underlying feelings.

Psychiatrists now command a formidable arsenal of psychotropic drugs, which can be of great benefit in treating mental illness.

- *Major tranquilizers* are chiefly *antipsychotic* drugs used to manage the symptoms of schizophrenia.
- *Minor tranquilizers* have much less dramatic effects than the major tranquilizers and are used on an outpatient basis to help manage anxiety disorders.
- *Antidepressants* are used to alleviate major depression.
- *Lithium (lithium carbonate)* is used to control bipolar disorder—though its primary effect is to reduce the manic phase more than the depressive phase of this disorder.
- *Ritalin* is widely prescribed for the learning disorder known as *attention deficit disorder* (ADD) in children.

While drugs such as these have doubtless significantly improved the lot of the mentally ill—especially those with severe illnesses—they are not without serious side effects, and they are perhaps too often used in lieu of counseling and other therapies directed at the causes of the disorders in question.

Psychiatric surgery is another approach to treating certain kinds of mental illness. Archaeologists have found evidence of primitive forms of such surgery in skulls that have been surgically opened through *trephining,* presumably to let out evil spirits. While various surgical techniques have been developed on an experimental basis, the most widely used, prior to the introduction of the major tranquilizers, was the *prefrontal lobotomy,* in which the brain's frontal lobe was permanently severed from the rest of the brain. The desired outcome of this drastic surgery was to control intractably combative schizophrenics. Indeed, many patients so treated did become calm —to the point of total unresponsiveness. In some cases, however, the surgery had precisely the opposite of the desired effect, actually aggravating violent behavior. The procedure is not used today.

Another medical intervention introduced before the era of effective psychotropic drugs was *electroconvulsive therapy* (ECT), popularly called "shock treatment." ECT sends a high-voltage, low-current charge through the brain of a sedated patient. The desired outcome is alleviation of depression, and, in many cases, ECT does achieve this outcome. However, side effects may include memory loss. ECT is still used today to treat some cases of severe depression that are unresponsive to drug therapy.

The counseling and analytical approaches of clinical psychologists typically reflect the theoretical orientation of the therapist, although many therapists today take an eclectic approach, borrowing from various approaches to treatment.

- Psychoanalysis and other psychodynamic approaches require the therapist to work for an extended period with the patient in an effort to investigate the nature of inner conflicts underlying the mental illness. As Freud saw it, by making the conflict-causing material of the unconscious available to consciousness, insight could be achieved, and individuals, by understanding their conflicts, would be able to cope with them.
- Humanistic therapies include the *client-centered* approach of Carl Rogers and many other therapeutic techniques that are directed toward increasing the individual's awareness and understanding of self. The goal of such understanding is the successful integration of the various, often conflicting, elements of personality and, ultimately, the promotion of self-growth.
- Cognitive therapies, as advanced by Aaron Beck, Albert Ellis, and others, focus on understanding (and altering) the thinking behind abnormal or maladaptive behavior.

Finally, some psychologists altogether avoid analysis of the causes, thought processes, or emotions involved in mental illness and address instead the abnormal or maladaptive *behavior* associated with the disorder. These *behavior therapies* attempt to adapt the principles of conditioning (Chapter 12) to the therapeutic situation. Behaviorist approaches are most effective in dealing with disorders associated with specific behaviors, especially substance abuse, compulsive gambling, enuresis (bed-wetting), and phobic disorders. In these situations, the abnormal behavior can be readily identified and *modified* through the application of reinforcement and other conditioning techniques.

CHAPTER 17

THE PSYCHOLOGY OF GROUPS

S*OCIAL PSYCHOLOGY* IS THE PSYCHOLOGY OF GROUPS: HOW INDIVIDUALS INFLUENCE and are in turn influenced by others. Although allied with sociology, social psychology focuses chiefly on the *individual* in relation to society, rather than on society and social structures. Generally speaking, the focus of social psychology is cognitive, and while social psychology is more firmly rooted in psychology than in sociology, it breaks with much of mainstream psychology in its situational approach to the important issue of personality. Most social psychologists view the individual as less autonomous and less defined by a stable personality than by particular relations to other people in particular situations.

Principles

From the perspective of the individual, the foundation of social psychology consists in identifying and studying five elements:

1. *Beliefs:* What we hold to be true about people and things. Typically, beliefs may be stated as propositions: "Most people are good."
2. *Attitudes:* These are evaluative beliefs—likes and dislikes. The issue of attitudes—where they come from and how they influence behavior—is a key focus of social psychology.
3. *Schemas:* Social psychologists think of schemas in much the same way as Piaget thought of them (Chapter 10)—as the ways we organize knowledge

and understanding, with new situations either being assimilated into our schemas or causing us to accommodate our schemas (extending them or creating new ones) to them. For the social psychologist, our schemas are made up of knowledge, beliefs, and attitudes.

Let us pause here to discuss the function of schemas a bit further. Social psychologists are particularly interested in how schemas generate errors, exaggerations, and other misconceptions about other people and groups. This is the result of *schematic processing,* by which people are simply assimilated into our schemas. That is, we respond to our schema rather than to the individual person before us. The most common effects of schematic processing include:

- *Stereotyping:* Ascribing certain traits or characteristics to someone on the basis of a schema rather than on direct knowledge of or experience with the individual.
- *Prejudice:* The negative beliefs and attitudes that may be associated with stereotyping.
- *Discrimination:* Taking action in accordance with prejudice.

We move on to the fourth and fifth basic elements of social psychology:

4. *Impressions:* The attitudes we form toward a person. Impressions may be products of the elements just discussed, in addition to physical, nonverbal, and verbal *information* gathered from individuals themselves.

Impression formation is an important area of study for the social psychologist. Not only are the roles of schemas and other information important in the study of impression formation, but so is the *order of information:* what we notice first about a person. Social psychology lends a great deal of scientific weight to what our mothers told us: *first impressions are very important.*

5. *Attributions:* Our assessment of the motives of others.

The area of attributions is rich and fairly complex. Social psychologists view attributions as having two dimensions:

1. *Stable* versus *unstable:* We attribute behavior either to relatively enduring (stable) or relatively changeable (unstable) motives.
2. *Internal* versus *external:* We attribute behavior to motives within people or outside of them.

These two dimensions produce a table of potential causes for a person's behavior:

- *Internal-stable:* These attributions encompass personality traits, intellectual capabilities, and fixed motives.
- *External-stable:* These attributions include enduring environmental factors

that help or hinder, such as the level of difficulty of something we attempt.

- *Internal-unstable:* These attributions include moods and effort.
- *External-unstable:* These attributions include changeable environmental factors, such as luck and opportunity.

Correspondent-inference theory, the classic explanation of attributions, identifies the following phenomena:

- *Fundamental attribution error:* Describes a tendency to attribute behavior exclusively to internal-stable causes, automatically excluding all other causes. That this error is pervasive suggests that people prefer to think of personality as stable and enduring.
- *The discounting principle:* We tend to make attributions based on whatever possible causes happen to catch our attention. Those that don't capture our attention we tend to discount.
- *The social desirability effect:* We tend to give more weight to socially negative motives than to socially positive ones. If a new acquaintance does you a favor, your tendency is to believe that this person is acting in the expectation of putting you in his or her debt—not out of mere generosity.
- *The actor-observer effect:* We tend to attribute the behavior of others to internal causes, while we tend to interpret our own behavior as the result of external causes, especially in situations that have negative outcomes. That Joe Blow received a grade of D is due to his laziness: he didn't study. That I received a grade of D is due to the difficulty of the course material and the unfairness of the instructor.
- *Gender biases:* These often play an important role in attributions. For example, a male executive who displays leadership initiative may be called "ambitious," while a female executive who behaves the same way may be identified as "pushy."

Attitude Change

Thus far we have emphasized how our schemas shape the attributions we make; obviously, however, our attitudes can also change. The social psychologist investigates precisely how and why they change.

Persuasive Appeals

Persuasive appeals are attempts to change attitudes by direct means. The effectiveness of a persuasive appeal, according to Carl Hovland, depends on four categories of characteristics:

1. *Communicator characteristics:* Attributes that make the communicator seem credible. These may include attractiveness, reputation, and straightforward body language.
2. *Message characteristics:* Does the message make logical sense? Is it clearly presented? (Other characteristics of an effective message include subtle use of repetition and emphasis.)
3. *Channel characteristics:* Through what medium is the message presented? The same message delivered on a national network television show may be more persuasive than if it were delivered on a street corner or even on a local radio broadcast.
4. *Audience characteristics:* The message and its presentation are most persuasive if they are geared to the audience. A sophisticated, well-informed audience will be more readily persuaded by a well-argued presentation that presents all sides. A less sophisticated, relatively uninformed audience will be more effectively moved by a simple presentation that focuses on one side only.

Indirect Causes of Change

Attitude change is not always the result of direct persuasive appeals. In the 1950s, Leon Festinger proposed the *cognitive dissonance theory,* which holds that we naturally tend to try to reconcile differences or conflicts between our attitudes and between our attitudes and our behavior. An important point to note is that, whereas Hovland assumed that behavior follows attitude, Festinger suggested that cognitive dissonance is resolved by first changing behavior, with attitude following. Festinger demonstrated this with a classic experiment:

> A subject is asked to perform a boring task. After he does this, the subject is then asked to persuade another subject (who is actually a *confederate* of the experimenter) to perform the task by persuading him that it is interesting and fun. That is, the experimenter has asked the subject to lie. Festinger paid some of his subjects a very small sum in return for lying, and he paid others substantially more. The subjects who were paid a small sum later reported that the task was very interesting after all, while those who were paid more substantially for lying did not change their attitude. They still thought the task was boring.
>
> Festinger concluded from this experiment that the subjects who had lied in return for a paltry sum felt compelled to justify what they had done—to resolve their cognitive dissonance by changing their attitude toward the task *after* the behavior of lying. Those who had been paid more substantially for

> lying felt that the money justified the lie and, therefore, did not feel compelled to change their attitude about the task.

Conformity

Attitude (as well as behavior and judgment) may also change in *conformity* with the perceived attitude of others. The classic experiment demonstrating this was conducted by Solomon Asch, who observed the behavior of a subject placed among a group of the experimenter's confederates who, as a group, purposely made erroneous judgments. More often than not, the subject went along with these group judgments. We see conformity in many aspects of life, ranging from issues of politeness, to fashion and fads, to sometimes menacing political beliefs (much has been made of the role of conformity in the German Nazi movement of the 1930s and 1940s, for example).

Compliance

Compliance differs from conformity in that it is a function of behavior rather than attitude. To *conform* is to shape one's attitude to fit that of the group, whereas to *comply* is merely to behave in a way that someone wants one to behave in.

Compliance is usually easier to obtain than conformity, and in many situations, people can get what they want from another person by obtaining compliance rather than conformity. Sales professionals have long been familiar with the *foot-in-the-door technique:* you ask your prospect for a little at first, then gradually ask for more. Another approach is the *door-in-the-face technique,* in which you ask for a lot, are refused, then "settle for" less. This is a common technique in negotiation, which plays upon the *norm of reciprocity:* a tacit belief that when someone gives you something (in this case, apparently makes a concession), you owe him something in return.

Sales professionals also practice the *that's-not-all technique,* offering the potential customer something at a certain price, then "throwing in" extras "for free." Persuaded by the apparent rise in value offered, you accept the original price.

Obedience

Beyond compliance is *obedience.* Compliance involves going along with something without necessarily changing your attitude. Obedience is doing what you are told regardless of how you feel. The classic experiment concerning obedience was conducted by Stanley Milgram in the 1960s:

> A subject who is designated the "teacher" is seated before an array of switches labeled with progressively higher levels of electric shock. The experimenter firmly instructs the teacher to read words into a micro-

phone to a "learner" seated in the next room. The experimenter explains to the teacher that the learner is wired to an apparatus that will deliver shocks controlled by the teacher. The experimenter firmly instructs the teacher to administer progressively stronger shocks each time the learner makes a mistake recalling the words read to him. (The learner is, in fact, a confederate, and no shocks are actually given—but the teacher doesn't know this, of course.)

Milgram found that about two-thirds of the "teachers" (the subjects) obeyed the experimenter's instructions, administering progressively higher shocks despite the learner's (feigned) cries of pain—and even after the learner mentioned that he had a heart condition! When a teacher showed unwillingness to continue, the experimenter *commanded* that the teacher continue. However reluctantly, most of the subjects indeed continued.

Milgram introduced additional variables into the experiment. If the teacher could see the learner through the window, obedience was more difficult to obtain. If the learner was in the same room, obedience was even more difficult to elicit. The presence of a second experimenter who vocally disagreed with the commands of the first experimenter also reduced the extent of obedience.

Observational Learning

Milgram's experiment was controversial, and his findings were (and remain) disturbing. Equally provocative has been research into the relationship between *aggression* and *observational learning.* While Freud (and others) believed that aggression is an innate human trait, many social psychologists espouse the *social learning* view, which holds that aggressive behavior is primarily learned. (Even if the *potential* for aggression is assumed to be innate, they argue, aggressive *behavior* is largely learned.) Here the classic experiment was conducted by Albert Bandura:

Bandura had three groups of young children observe an adult play with a Bobo doll (a life-size inflatable clown figure weighted at the bottom, so that it returns to an upright position when punched). The adult hit, punched, and kicked the Bobo doll. One group of children saw the adult praised for this activity. Another group saw the adult punished for this activity. The third group saw neutral consequences: the adult was neither praised nor punished.

> After this phase of the experiment, Bandura found that the children who saw the adult being praised (reinforced) were the most aggressive of the three groups when it became their turn to beat up the Bobo doll. Moreover, they tended to emulate the adult's behavior more closely than the other two groups.
>
> Later, all three groups were offered incentives for imitating the adult's battery of the Bobo doll. At this, all three groups thoroughly emulated the adult's actions. There were no significant differences between the groups.

Bandura's work has often been cited by opponents of television violence, who argue that exposing children to images of violent behavior enhances their aggressive behavior—and, ultimately, creates a more violent society.

PART THREE

MIDTERMS AND FINALS

CHAPTER 18

CARTHAGE COLLEGE

PSYCHOLOGY 110: GENERAL PSYCHOLOGY

Jeffrey A. Gibbons, Visiting Assistant Professor

TWO MULTIPLE-CHOICE EXAMS ARE GIVEN IN THE COURSE, WHICH, TOGETHER, account for about half of the course grade. I also give frequent quizzes in class. Sometimes I ask poorly worded questions or circle the wrong answer to challenge them to think. If they do not challenge me, I either move on or prompt them to challenge me (i.e., ask them if they have a problem with specific questions or answers).

My primary objectives in teaching this course are

1. To get my students to think and then think critically
2. To get my students interested in psychology
3. To challenge my students and make them better students

As a result of taking Psychology 110, students

1. Should be able to define and provide examples for terms such as *independent* and *dependent variables*
2. Should have knowledge of basic physiology
3. Should know that development is primarily continuous with periods of upheaval
4. Should know that experience creates top-down processing (and know that the following terms are related to top-down processing: *schemas, theories, concepts*), which drives perception, memory, and behavior
5. Should know that memory is malleable
6. Should know that language and thought are intertwined

7. Should know that IQ differences can be predisposed, but they are primarily created through their environment
8. Should know that life is predominantly pleasant
9. Should know that emotion theories are not necessarily antagonistic
10. Should understand the Freudian, trait, humanist, and learning theories for personality
11. Should be aware of the bystander effect and experiments on conformity and obedience.

MIDTERM EXAM

Most of the multiple-choice questions can be easily answered if students can define and provide examples for each term in the course textbook. For the remaining questions, they should be able to synthesize information from different chapters of the book.

1. Physical changes in adulthood include
   a. decreased sensory abilities.
   b. decreased strength after 23.
   c. increased stamina.
   d. all of the above

Answer: a

2. Which theory or theories is or are correctly explained?
   a. The trichromatic theory involves three color receptors.
   b. Firing rate of neurons matching pitch is frequency theory.
   c. Place theory refers to different pitches affecting the cochlea in different places.
   d. all of the above

Answer: d

If you are *totally* in the dark about a question and need to guess, remember that—*more often than not*—an "all of the above" multiple-choice response is correct.

3. Which of these statements is true?
   a. The myelin sheath slows the action potential.
   b. The axon releases neurotransmitters into the synapse.
   c. Neurotransmitter is to receptor as key is to lock.
   d. all of the above

Answer: c

4. Mental changes in adulthood include

a. decreased ability to recall names.
b. increased ability to reason quickly (math and logic).
c. decreased facial recognition.
d. decreased vocabulary.

Answer: a

5. How do adolescents develop physically?

a. Primary sex characteristics, such as ovaries and testes, develop.
b. The first ejaculation occurs during a dream for most men.
c. Secondary sex characteristics occur, such as the development of pubic hair, hips, and breasts.
d. all of the above

Answer: d

6. Which of these statements is true?

a. Flashbulb memories are always remembered correctly because they are created from emotional events.
b. Explicit memory refers to automatic processes such as shifting gears in a car.
c. Short-term memory holds memories for a lifetime.
d. Iconic and echoic memory refer to vision and audition, respectively.

Answer: d

In eliminating incorrect multiple-choice responses, look for those that include absolute statements, such as "*always* remembered" in choice a. Absolute statements are often false.

7. Which of these statements is true for neurons?

a. The threshold is the minimum amount of stimulation necessary for a neuron to fire.
b. A terminal bouton holds vesicles containing neurotransmitters.
c. Synapses are gaps between neurons and other neurons, or gaps between neurons and muscles.
d. all of the above

Answer: d

8. Which of these memory techniques is (are) true?

a. Chunking is a method of organization where the few important concepts are introduced first, followed by the many less important concepts.

b. Hierarchies put information in groups.

c. The method of loci is a mnemonic using location, association, imagery, and order to aid memory.

d. all of the above

Answer: c

9. Which of these statements is true?

a. *Accommodate* means to incorporate new information into the schema.

b. *Assimilate* means to change the schema.

c. Jean Piaget studied adult development in stages.

d. Schemas are concepts that organize information.

Answer: d

Note that choices a and b are reversed; that is *assimilate* means to incorporate new information into a preexisting schema, whereas *accommodate* requires changing a preexisting schema.

10. In an experimental study of the effects of anxiety on self-esteem, anxiety would be the ________ variable.

a. extraneous

b. independent

c. dependent

d. correlational

Answer: b

The *independent variable* is what the researcher is studying, whereas the *dependent* variable is what the researcher measures to determine if the independent variable has an effect on it.

11. Authoritative parenting styles

a. include setting and explaining rules and backing them up.

b. inhibit self-esteem, self-reliance, and social competence.

c. include setting rules without explanation.

d. a and b

Answer: a

All the exam questions are equally important. I tell students that I could test them completely on one textbook chapter, which makes the point that they need to know all the information and they'd better be able to define and provide examples for each term.

12. Psychologists who emphasize the impact of learning on development are likely to favor perspectives that emphasize both

a. nurture and stages.
b. nature and continuity.
c. nurture and continuity.
d. nature and stages.

Answer: c

Choice c is best because it mentions factors most directly relevant to learning.

13. In an experimental study of the effects of sleep deprivation (varying sleep) on memory, memory would be the

a. control condition.
b. independent variable.
c. experimental condition.
d. dependent variable.

Answer: d

The *dependent variable* is what the researcher measures. It is the effect, whereas the *independent variable* is the cause.

14. Which of these statements about experimentation is true?

a. Extraneous variables are "outside influences," which are controlled in a thorough experiment.
b. A population is a small group of people that we test.
c. A sample is a large group of people that we study.
d. all of the above

Answer: a

15. Which of these statements is true for sensation and perception?

a. Subliminal messages can influence our behavior in subtle ways.
b. Absolute threshold is the minimal amount of stimulation to sense/perceive a stimulus.
c. Signal detection refers to how perception of signals can be guided by psychological states, consequences for mistakes, stimuli acting as "background noise" (interference), and stimulus strength.
d. all of the above

Answer: d

16. Which of these statements is true about memory?

a. Leading questions can change memory.
b. The cognitive interview uses open-ended questions and imagery to improve memory.
c. Memory decreases over time.
d. all of the above

Answer: d

17. Which statement is true?

a. Touch receptors sense the physical shapes of particles.
b. Smell is a mix of pressure, warmth, cold, and pain.
c. Taste receptors sense sweet, salty, sour, and bitter.
d. Gate-control theory states that the brain blocks or allows pain.

Answer: c

18. Compared to clinical psychologists, most psychiatrists are especially likely to adopt a _______ perspective.

a. behavioral
b. biological
c. cognitive
d. social-cultural

Answer: b

Remember that psychiatrists are physicians—*medical* doctors.

19. Jennifer proofreads manuscripts for a publisher and is paid $10 for every three pages she reads. Jennifer is reinforced on a _______ schedule.

a. fixed-interval
b. variable-interval
c. fixed-ratio
d. variable interval

Answer: c

Look for the answer in the question. Even if your memory doesn't serve you well here, it is not difficult to conclude that a set fee paid for a set number of pages is a *fixed* ratio.

20. A theory

a. is composed of practical and testable predictions.
b. organizes facts.
c. seems a lot like a schema.
d. all of the above

Answer: d

Students have approximately one minute to answer each question. I tell them to move on to other questions if they feel uncomfortable (anxious) about a question. I also tell them to recheck their answers, but only to make changes if they are absolutely sure of the new response.

21. The constriction of your pupil

a. is a voluntary response when someone shines a light in your eye.
b. involves sensory neurons only.
c. is a conditioned response shaped through punishment.
d. following the sound of a bell *without light* is most likely a conditioned response created through Pavlovian (classical) conditioning.

Answer: d

Many students may find that the best way to answer this question is to eliminate the wrong answers.

22. Research indicates that memories retrieved during hypnosis are:

a. forgotten again as soon as the person awakens from the hypnotic state.
b. often a combination of fact and fiction.
c. accurate recollections of information previously learned.
d. wholly products of the imagination.

Answer: b

23. Which of these statements is true?

a. The same sperm and egg create identical twins.
b. Different eggs and sperms create fraternal twins.
c. The environment and genetics together explain behavior.
d. all of the above

Answer: d

24. Which of these statements is true?
a. Relearning happens more slowly than initial learning.
b. Recall improves with time and age regression.
c. Recognition is noticing that things are familiar.
d. Priming is the deactivation of memory.
Answer: c

25. Stage 4 sleep is to sleepwalking as stage 1 sleep is to
a. hallucinations.
b. night terrors.
c. sleep talking.
d. alpha waves.
Answer: a

26. If college graduates typically earn more money than high school graduates, this would indicate that level of education and income are
a. causally related.
b. positively correlated.
c. independent variables.
d. negatively correlated.
Answer: b

Two variables are positively correlated if they consistently change together in the same direction. While positive correlation suggests cause and effect, it is not proof of cause and effect.

27. Which of these statements is true about drugs?
a. Withdrawal occurs when drug use *starts.*
b. Bodily need of drugs is psychological dependence.
c. Stimulants decrease neural activity.
d. Tolerance increases with drug use.
Answer: d

28. Which of these statements is true?
a. Denial and acceptance are stages for the terminally ill.
b. No evidence exists for the midlife crisis.
c. The social clock relates to times for adult commitments.
d. all of the above
Answer: d

29. If two objects are assumed to be the same size, the object that casts the smaller retinal image is perceived to be

a. closer.
b. more distant.
c. more coarsely textured.
d. less hazy.

Answer: b

30. A child's fear at the sight of a hypodermic needle is a(n)

a. conditioned stimulus.
b. unconditioned response.
c. conditioned response.
d. unconditioned stimulus.

Answer: c

This question requires an assumption that the child has had experience with vaccination.

31. Remembering how to solve a jigsaw puzzle without any conscious recollection that one can do so best illustrates ________ memory.

a. short-term
b. explicit
c. flashbulb
d. implicit

Answer: d

If you don't know the answer, think about the meaning of the words. What is the difference between *explicit* and *implicit?*

32. A segment of DNA capable of synthesizing a specific protein is called a(n)

a. neuron.
b. teratogen.
c. interneuron.
d. gene.

Answer: d

33. Which of these statements is true?

a. Fertilization is the splitting of the sperm and the egg.
b. The ovum and sperm are organs producing reproductive cells.
c. Testosterone is a male hormone.
d. all of the above

Answer: c

34. Which statement is true of neurotransmitters and drugs?

a. An antagonist binds and blocks the effects of a neurotransmitter.
b. An agonist binds and blocks the effects of a neurotransmitter.
c. The blood-brain barrier allows all chemicals to pass.
d. all of the above

Answer: a

35. A case study is a research method in which

a. a representative sample of people are questioned regarding their opinions or behaviors.
b. an individual is studied in great detail.
c. organisms are carefully observed in their natural environment.
d. an investigator manipulates one or more variables that might affect behavior.

Answer: b

36. Attachment

a. is created by attention from parents and forms basic trust.
b. can be either secure or insecure.
c. is created by familiarity and body contact.
d. all of the above

Answer: d

37. Which of these statements is true?

a. Rewards and punishment are important for Piaget's preoperational stage and Kohlberg's preconventional level.
b. Moral action is when talk and action are different.
c. Adolescents communicate about sex and are educated about sex.
d. Research shows that children are happier than adults.

Answer: a

It helps to remember that Kohlberg's work is closely related to that of Piaget.

38. The personal values of psychologists are likely to influence their choice of

a. topics of investigation.
b. research methods.
c. explanatory theories.
d. all of the above

Answer: d

39. Punishment is a potentially hazardous way for parents to control their young children's behaviors because

a. punishment cannot even temporarily restrain undesirable behaviors.
b. the use of punishment could condition children to fear and avoid their parents.
c. the more severely children are punished for undesirable behaviors, the more likely they will exhibit those behaviors.
d. children will forget how to perform punished behaviors even when they are justified and necessary.

Answer: b

40. How is electromagnetic energy (light) transformed and then transduced?

a. Neurons take the message from bipolar cells to ganglion cells to the optic nerve to the visual cortex.
b. The rods and cones transduce light and color, respectively.
c. The pupil allows light in and the lens flips and focuses the image.
d. all of the above

Answer: d

41. Research comparing parental care in the home with professional day-care programs outside the home indicates that

a. children who receive professional day care actually spend more quality time each day with their parents.
b. professional day care is more appropriate for infants than for older preschool children.
c. the quality of child care is more important than whether it is provided inside or outside the home.
d. a and b

Answer: c

42. Which statement or statements relate(s) to how sound (changes in air pressure) is transduced?

a. The hammer, anvil, and stirrup send vibrational energy to the cochlea.
b. Air pressure vibrates the eardrum, which vibrates the bones of the middle ear.
c. From vibrations on the cochlea, fluid moves hair cells, which cause neurons to fire.
d. all of the above

Answer: d

Some questions have more than one answer. For these questions, the student should choose the best answer, which may entail an "all of the above" response. I tell them to look for the critical concept in each answer or response. Then they should look away from the exam and define the term. Only then should they look for the answer or try to match their response to the question. In essence, they are choosing the best answer by eliminating incorrect responses.

43. Which statement is true for conditioning?

a. Generalization is responding differently to similar stimuli.
b. Habituation is increased responding due to constant stimulation.
c. It is best if unconditioned stimuli precede neutral stimuli.
d. Extinction is the reduction of a conditioned response.

Answer: d

44. During the course of successful prenatal development, a human organism begins as a(n) _________ and finally develops into a(n) _________ .

a. embryo; zygote
b. zygote; fetus
c. embryo; fetus
d. zygote; embryo

Answer: b

45. Which of these statements about brain scans is true?

a. The PET scan creates a picture of the brain from electrical activity in the brain.
b. The EEG creates pictures/graphs of brain activity from the use of radioactive glucose in the brain.
c. The CAT scan creates a picture of the brain from X rays of the brain.
d. all of the above

Answer: c

Note that a and b are reversed; a PET scan uses radioactive tracers, whereas an EEG measures electrical activity.

46. The rooting reflex refers to a baby's tendency to

a. withdraw a limb to escape pain.
b. open the mouth in search of a nipple when touched on the cheek.
c. be startled by a loud noise.
d. look longer at human faces than at inanimate objects.

Answer: b

47. Which statement is true about reinforcement?

a. Negative reinforcement decreases responding.
b. Shaping uses reinforcers to mold new responses from old responses.
c. A primary reinforcer needs to be learned.
d. Delayed reinforcers occur soon (seconds) after a response.

Answer: b

48. Which of these statements is true?

a. Middle items are remembered poorly in the serial position effect.
b. Rehearsal decreases memory retention.
c. Effortful processes happen quicker than automatic processes.
d. Automatic processing requires a great deal of effort.

Answer: a

Incorrect responses are often incorrect because they simply make no sense. Such is the case here. Don't depend on rote recollection of information. Read critically and *think*.

49. Which statement is true for memory?

a. Freud believed that memory faded quickly but Ebbinghaus believed all memories were stored.
b. Memory is distributed all over the brain.
c. Memory is stored in the hippocampus.
d. a and b

Answer: b

50. Narcolepsy is a disorder involving

a. difficulty falling and staying asleep.
b. the temporary cessation of breathing during sleep.
c. severe fright which occurs during stage 4 sleep.
d. periodic uncontrollable attacks of overwhelming sleepiness.

Answer: d

FINAL EXAM

Take the following steps to achieve success in this course:

1. **Read.**
2. **Test yourself by defining terms and providing examples.**
3. **Think.**
4. **Challenge your instructor and other students inside and outside class in an intellectual manner.**
5. **Be aggressive in the classroom—because you are paying a great deal of money to take tests you could take over the Internet.**

You should also at least look at the terms before coming to class each day. Form questions about the material when you are confused or when you have an "interesting" thought. Ask your question or make your comment in class to help the class and yourself. You should try to find fault with your instructor's knowledge, which will keep everyone on their toes. Think critically, and ask probing questions, even when doing so may seem "stupid." Test each other before the midterm and final, and come to review sessions even when you "already know everything"—maybe *especially* then!

1. Which statement is true?

a. Chomsky and Skinner advocated environment and genetics respectively.
b. Babbling is learned one-word phrases.
c. Grammar includes rules about meaning (semantics) and word order (syntax).
d. One-word phrases include noun/verb or noun/adjective pairs.

Answer: c

2. In considering gender differences, it is important to remember that

a. we are often more fascinated by differences than by similarities between genders.
b. gender stereotypes do not always reflect actual gender differences.
c. variations within the genders typically exceed variations between the genders.
d. all of the above

Answer: d

3. Men tend to have or show more

a. successful suicides.
b. color blindness.
c. alcoholism and antisocial behavior.
d. all of the above

Answer: d

4. Which statement is true?

a. The id mediates the ego and the superego.
b. The superego is the selfish entity that drives our behavior.
c. The ego is the moralistic part of personality.
d. none of the above

Answer: d

5. Which statement is true?

a. Fixation is inflexibility toward solving a problem.
b. McGyver has functional fixedness.
c. Mental set is trying new solutions.
d. Confirmation bias proves a theory wrong.

Answer: a

6. Assuming everyone in a group is the same is

a. in-group bias.
b. scapegoating.
c. prejudice.
d. stereotyping.

Answer: d

7. Which statement is true?

a. Free association is saying the first words that come to mind.
b. The TAT allows people to project their beliefs into a picture by writing about it.
c. Psychoanalysts try to help clients uncover themes of dreams.
d. all of the above

Answer: d

8. As we age, men and women become

a. increasingly heterosexual.
b. less similar.
c. increasingly homosexual.
d. none of the above

Answer: d

9. Health is increased best by

a. lifting heavy weights (anaerobic exercise).
b. watching newscasts describing crime scenes.
c. exercising aerobically and eating many small meals.
d. eating snacky cakes and cheesy poofs.

Answer: c

10. Which statement is true?
- a. Set point is the weight at which one's body wants to be.
- b. Anorexia is self-controlled starvation.
- c. Metabolism is the rate your body uses glucose.
- d. all of the above

Answer: d

11. Emotions are composed of arousal and thinking. This theory is related to passionate love and is called the
- a. Opponent process theory.
- b. James-Lange theory.
- c. Cannon-Bard theory.
- d. Two-factor theory.

Answer: d

12. Which statement is true?
- a. Intelligence may be shown at school, on the street, on the court, or on a canvas.
- b. Creativity entails imagination, taking risks, knowledge, and mental stimulation.
- c. Japanese score highest on IQ tests followed by Caucasians and then African Americans.
- d. all of the above

Answer: d

13. Which statement is true?
- a. Dispositions are extrinsic.
- b. Situations are extrinsic.
- c. The Fundamental Attribution Error states that people overestimate situations.
- d. a and c

Answer: b

14. Believing that life is fair for everyone because your life is good is called
- a. the feel-good, do-good phenomenon.
- b. the just-world phenomenon.
- c. dissonance.
- d. none of the above

Answer: b

15. Twelve-year-old Jerry has an IQ of 75 on the original version of the Stanford-Binet intelligence test. His mental age is:

a. 8
b. 9
c. 12
d. 16

Answer: b

> Note that 9 is 75 percent of 12. In the *original version* of the IQ test, an IQ of 100 is precisely at age level. This question requires knowledge as well as elementary mathematical ability.

16. Behavior therapies

a. use reinforcers for autism and child rearing.
b. punish inappropriate responses such as drinking or bed-wetting.
c. replace fear responses with being calm.
d. all of the above

Answer: d

17. Common ancestors and cultural heritage are

a. collectivism.
b. ethnicity.
c. personal space.
d. individualism.

Answer: b

18. Which statement is true?

a. Standardization means using a control group for comparison.
b. Validity is easily measured for IQ tests.
c. Normal distributions are frequency curves where most of the scores are at the ends.
d. Reliability means measuring what you are supposed to measure.

Answer: a

19. Good leaders

a. use social and task leadership as necessary.
b. recognize (say "good job") and challenge their subordinates when appropriate.
c. compliment employees in public and chastise employees in private.
d. all of the above

Answer: d

20. Which statement is true?
a. Regression is retreating to an earlier age.
b. Displacement is accusing another of your feelings.
c. Blocking thoughts or feelings is sublimation.
d. Doing the opposite of impulses is rationalization.

Answer: a

Each question typically tests a group of related concepts. Students must know each concept by being able to define it and provide an example of the concept. Otherwise, they will be fooled on the questions and make a mistake.

21. Which statement is true?
a. Passionate love relates to friendship mostly, not attraction.
b. Getting involved is the bystander effect.
c. Social exchange theory involves weighing outcomes before behaving.
d. Attraction cannot occur through exposure (familiarity).

Answer: c

22. A person suffering from a phobia could receive best treatment from
a. a psychoanalyst.
b. a humanistic therapist.
c. a behavior therapist.
d. a drug dealer on the street.

Answer: c

Potentially a difficult question (choosing from among a, b, and c), but recall the nature of phobia as strongly related to issues of stimulus and response.

23. Which statement is true?
a. The same number of phonemes is used by each language.
b. Morphemes are the smallest units of sound that possess meaning.
c. Opposite beliefs align when united.
d. Our beliefs should be involved with decision making because they direct logic.

Answer: b

24. Psychotherapy is most likely to be effective in freeing:
a. Sharon from the feeling that her life is meaningless and worthless.
b. Jim from an excessive fear of giving speeches in public.
c. Portia from her delusions of persecution and auditory hallucinations.
d. Luther from his antisocial personality disorder.

Answer: a

The correct response is based on an understanding of the uses and limitations of psychotherapy versus other forms of treatment, including medication and behavior modification.

25. Noticing physiology influences emotions. This is the
a. opponent process theory.
b. James-Lange theory.
c. Cannon-Bard theory.
d. Two-factor theory.
Answer: b

26. Which statement is true?
a. Fixating in the oral stage may entail the inability to have sex.
b. The genital stage is where sexual feelings are repressed.
c. Fixating in the anal stage may relate to obsessive-compulsive behaviors.
d. all of the above
Answer: c

27. A stroke (coronary heart disease) is more likely to happen
a. for type A personalities than type B personalities.
b. when daily hassles become overwhelming.
c. to low blood pressure sufferers than to high blood pressure sufferers.
d. a and b
Answer: d

28. Which statement is true regarding a polygraph?
a. A polygraph works for everyone, every time.
b. A polygraph measures a variety of physiological functions.
c. Polygraphs are of no use in criminal investigation.
d. Polygraphs work better for innocent people.
Answer: b

29. Which statement is true?
a. Aggression is affected by genetics, alcohol, hormones, TV, and parental behavior.
b. Attractive people are perceived as happier and more sensitive and dishonest.
c. Attractive people do not believe others are sincere.
d. all of the above
Answer: d

30. After having one emotion, we are soon sure to have the opposite emotion. This is the

a. Opponent process theory.
b. James-Lange theory.
c. Cannon-Bard theory.
d. Two-factor theory.

Answer: a

Think about the meaning of the word *opponent.*

31. According to Milgram, the most fundamental lesson to be learned from his study of obedience is that

a. people are naturally predisposed to be hostile and aggressive.
b. even ordinary people, who are not usually hostile, can become agents of destruction.
c. the desire to be accepted by others is one of the strongest human motives.
d. people value their freedom and resist being coerced to do something.

Answer: b

32. Attitudes

a. are beliefs or feelings.
b. change to fit behavior before behavior will change to fit attitudes.
c. are always in line with action in a moral action.
d. all of the above

Answer: d

33. Which statement is true?

a. Most tests are aptitude tests.
b. Achievement tests measure future ability.
c. Aptitude tests measure future ability.
d. Aptitude tests measure past learning.

Answer: c

34. The humanistic approach

a. attempts to change where their clients attribute success or blame.
b. asks people about their childhood sexuality.
c. alters the environment to control behavior.
d. uses listening and rephrasing to help people.

Answer: d

Humanistic psychotherapy is said to be "client-centered."

35. Which statement is true?

a. There are more types of positive emotions than negative emotions.
b. Facial expressions remain constant across cultures.
c. High arousal is good for complex tasks (e.g., debating world economics).
d. The right hemisphere is more active for positive emotions.

Answer: b

36. Which statement is true?

a. An introvert spends a lot of time socializing.
b. The humanistic approach focuses on people being and feeling good and self-actualizing.
c. Agreeable is being trusting and helpful.
d. b and c

Answer: d

37. Which statement is true?

a. Homosexuals are considered abnormal by the DSM-IV.
b. Children who are born first seem to achieve more than their siblings.
c. High achievers are only externally motivated.
d. all of the above

Answer: b

38. Concern for personal goals and attributes is

a. collectivism.
b. personal space.
c. individualism.
d. ethnicity.

Answer: c

39. A person blaming the instructor for scoring a D

a. has a personality disorder.
b. has externalized the locus of control.
c. has internalized the locus of control.
d. is histrionic.

Answer: b

40. Women tend to have or show more
- a. anxiety.
- b. depression.
- c. interdependence.
- d. all of the above

Answer: d

41. Cancer is related to a busy immune system caused by
- a. socializing.
- b. frequent exercise.
- c. continued stress.
- d. relaxing.

Answer: c

42. After losing, the tendency to blame a single player (e.g, kicker or quarterback) is called
- a. stereotyping.
- b. in-group bias.
- c. scapegoating.
- d. prejudice.

Answer: c

43. Which statement is true?
- a. Availability heuristic is using prototypes to make decisions.
- b. Representativeness heuristic is using statistical information to make decisions.
- c. The way a question is framed has no bearing on the response.
- d. all of the above

Answer: a

44. Which statement is true?
- a. Schizophrenia entails delusions and/or hallucinations as a result of too much dopamine.
- b. A hard blow to the head would most likely cause depression.
- c. A narcissistic individual would probably wear a lamp shade at a party.
- d. A person who forgets his or her identity and leaves home is bipolar.

Answer: a

45. Which statement is true?

a. Depending on your teammates or coworkers to do your work is social loafing.
b. Giving a well-rehearsed speech in front of a crowd is social facilitation.
c. People participating in the L.A. riots were probably experiencing deindividuation.
d. all of the above

Answer: d

46. The best example of a category of objects, events, or people is called a(n)

a. algorithm.
b. heuristic.
c. concept.
d. prototype.

Answer: d

47. Repeatedly washing your hands is to _______ as repeatedly thinking about your own death is to _________.

a. somatoform disorder; schizophrenia disorder
b. mania; depression
c. phobic disorder; hypochondriasis
d. compulsion; obsession

Answer: d

48. Psychoanalysis may entail

a. dream analysis.
b. free association.
c. interpretation.
d. all of the above

Answer: d

49. In a complete sexual response cycle, _____________ immediately precedes ___________.

a. The excitement phase; orgasm
b. the plateau phase; orgasm
c. the plateau phase; the excitement phase
d. the excitement phase; the resolution phase

Answer: b

50. Helping severely depressed people may entail

a. electroconvulsive shock therapy.
b. prescribing antidepressants.
c. psychotherapy.
d. all of the above

Answer: d

The material for a course is important, but the instructor is the critical element in the classroom. We have to entertain, challenge, disrupt, enrage, and force students to attend class, think, and ask questions. Our focus should be on creating better instructors rather than creating better texts and tests.

CHAPTER 19

CENTRAL MICHIGAN UNIVERSITY

PSYCHOLOGY 100: INTRODUCTION TO PSYCHOLOGY

Byron Gibson, Associate Professor

THIRTEEN WEEKLY QUIZZES AND ONE CUMULATIVE FINAL ARE GIVEN IN THIS COURSE. The final is included here. All quizzes and exams are multiple choice.

The primary goal of this course is to familiarize students with the wide variety of subdisciplines within the field, and to ensure that they know the *basic* findings in those subdisciplines. I want my students to be able to converse knowledgeably on the main findings within the subdisciplines. For example, after the course, they should be able to identify Piaget's stages and the main accomplishments in each stage. Another example: They should be able to identify the different theoretical approaches to the study of personality. They need to have a foundation of basic psychological knowledge.

My philosophy of introductory psychology is that it should provide the student with a foundation of psychological knowledge that can be applied in future classes. Some professors focus on issues such as critical thinking; however, I believe that before one can think critically on a given topic, one must first have a background of knowledge in the topic. Therefore, I use the introductory psychology course to help provide that foundation.

FINAL EXAM

An effective way to prepare for the quizzes and the final is to create flash cards with the important terms and concepts on one side and the definitions on the other. Students should go through them until they can get through an entire chapter's worth without an error. I emphasize the difference between college and high school, and how much more study time is necessary to succeed in college. I also have a specific weekly timetable I suggest students follow: quizzes are Thursday afternoon; after taking the quiz, I suggest they go to the course Web site and get the hints for the following week's quiz; make flash cards on Thursday night and Friday (this gives a chapter "preview"). Read the chapters on Saturday and Sunday. Study flash cards on Monday and Tuesday. Go over class notes on Wednesday. Thursday morning, you should study the flash cards until you can get them 100 percent correct.

1. With respect to research done in the nineteenth century on brain size and race, which of the following statements is accurate?
a. Even respected scientists like Broca and Binet believed that brain size would be correlated with intelligence.
b. The researchers at that time found differences in brain size between races.
c. The research of some individuals has been shown to be fraudulent.
d. All of the above statements are accurate.
e. b and c only
Answer: d

2. Which of the following is not one of the skills that begins to develop in the concrete operational stage?
a. pictorial representation
b. make-believe play
c. intentional, goal-directed behavior
d. None of the above begin in the concrete operational stage.
Answer: d

3. Which of the following disorders does not go with the others?
a. phobias
b. obsessive-compulsive disorders
c. multiple personality disorder
d. panic disorders
e. generalized anxiety
Answer: c

4. William James is most closely associated with which school of psychology?

a. structuralism
b. functionalism
c. behaviorism
d. humanism

Answer: b

5. Social learning theorists emphasize the importance of

a. learning through reinforcement.
b. learning through association.
c. learning through observation.
d. learning without reinforcement.
e. both a and b
f. both c and d

Answer: f

6. In second-order conditioning,

a. what was originally a CS now becomes a UCS.
b. what was originally a UCS now becomes a CS.
c. what was originally a CR now becomes a UCR.
d. what was originally a UCR now becomes a CR.

Answer: a

7. The Rorschach (inkblot) test

a. is a projective test.
b. is said to be sensitive to unconscious conflicts.
c. lacked validity in most studies.
d. all of the above

Answer: d

8. Which of the following statements regarding bipolar disorder is true?

a. Manic episodes are often unaccompanied by subsequent depressive episodes.
b. Manic episodes often include delusions regarding the importance of the thoughts that the manic person is experiencing.
c. There is no evidence that bipolar disorder has a significant genetic component.
d. Women are more likely to be diagnosed as having bipolar disorder.

Answer: b

9. A therapeutic technique in which the patient is encouraged to act out his or her relationships with the therapist, and is also urged to experience very strong emotions, to swear, kick, and scream—would be what kind of therapy?
a. psychoanalytic
b. existential
c. rational emotive
d. gestalt

Answer: b

10. A person with an internal locus of control
a. is less likely to be depressed in general.
b. is less likely to succeed in general.
c. is more likely to experience bipolar disorder.
d. is more likely to have suicidal thoughts.

Answer: a

11. Hallucinations in schizophrenia
a. are referred to as negative symptoms.
b. are more likely to be auditory than visual.
c. are typical of catatonic schizophrenics.
d. none of the above

Answer: b

12. The independent variable is the hypothesized _______; the dependent variable is the hypothesized _______.
a. cause; cause
b. effect; effect
c. cause; effect
d. effect; cause

Answer: c

13. Which of the following behaviors made by Kenneth Bianchi in attempting to portray himself as a multiple personality would not be considered a mistake on his part?
a. initially portraying himself as having only two personalities
b. having the second personality come out only under hypnosis
c. attempting to shake hands with a hallucination
d. using the name of a real person as his second personality

Answer: b

14. Lithium is best used to treat
a. bipolar disorder.
b. schizophrenia.
c. anxiety disorders.
d. dissociative disorders.
Answer: a

15. Tardive diskinesia sometimes results from
a. treatment for bipolar disorder.
b. treatment for schizophrenia.
c. treatment for anxiety disorders.
d. treatment for dissociative disorders.
Answer: b

16. As patients progress through psychotherapy they will often shift toward becoming (or having) a(n)
a. high sensation seeker.
b. low sensation seeker.
c. internal locus of control.
d. external locus of control.
Answer: c

17. If intelligence is highly heritable, this means
a. it cannot be changed.
b. much of it is due to environmental factors.
c. much of it is due to genetic factors.
d. a and c only
Answer: c

18. Superstitions can develop because of
a. variable-ratio reinforcement.
b. fixed-ratio reinforcement.
c. continuous reinforcement.
d. noncontingent reinforcement.
Answer: d

19. A child who doesn't conserve quantity does not have an appreciation of the concept of
a. reversibility.
b. accuracy.
c. size.
d. weight.
Answer: a

20. According to the humanistic movement, an individual's personality is largely determined by

a. self-actualizing tendencies.
b. forces in the environment.
c. striving for superiority.
d. forces in the unconscious.

Answer: a

21. _______ transmit information to other neurons; _______ receive information *from* other neurons.

a. Axons; synapses
b. Dendrites; axons
c. Synapses; dendrites
d. Axons; dendrites

Answer: d

22. The reappearance of a conditioned response after extinction and a period of rest is called

a. disinhibition.
b. reconditioning.
c. stimulus generalization.
d. spontaneous recovery.

Answer: d

23. Which of the following dimensions is not a component of hardiness?

a. commitment
b. change
c. control
d. challenge

Answer: b

24. On average, short-term memory can hold _______ pieces of information.

a. 7
b. 9
c. 11
d. 13

Answer: a

25. The myelin sheath helps to increase the _______ of neural impulses.
a. frequency
b. intensity
c. threshold
d. speed
Answer: d

26. You sustain an ankle injury while running from a mugger. You don't feel any pain until you are in the safety of the police station, a half hour later. Your body administered _______ to you.
a. amphetamines
b. endorphins
c. serotonin
d. morphine
Answer: b

27. Salivation as a reaction to lemon juice in the mouth would be a(n)
a. conditioned response
b. unconditioned response
c. conditioned stimulus
d. unconditioned stimulus
Answer: b

28. In classical conditioning, an unreinforced trial is one in which
a. the CR does not appear.
b. the CS does not appear.
c. the US is not presented.
d. the orienting action does not appear.
Answer: c

29. John has suffered brain damage in Broca's area. He will most likely have trouble
a. remembering past events.
b. speaking fluently.
c. staying awake.
d. understanding others when they speak.
Answer: b

31. The partial reinforcement effect means that extinction will occur more quickly after _______ reinforcement.

a. continuous
b. variable-interval
c. variable-ratio
d. fixed-ratio

Answer: a

32. Words, events, places, and emotions that trigger our memory of the past are called

a. encoding devices.
b. iconic traces.
c. context effects.
d. retrieval cues.

Answer: d

33. Which defense mechanism involves the conscious expression of feelings that are the opposite of unconscious feelings?

a. repression
b. reaction formation
c. sublimation
d. projection

Answer: b

34. Which of the following treatments is likely to lead to a slight memory loss?

a. aversion therapy
b. systematic desensitization
c. ECT
d. lithium treatment

Answer: c

35. The concept of a semantic network is important in understanding

a. memory.
b. defense mechanisms.
c. neurotransmitters.
d. aversion therapy.

Answer: a

37. The physiological hypothesis regarding introversion/extroversion suggests that introverts' brains are _______ compared to extroverts' brains.

a. overstimulated
b. understimulated
c. quicker
d. slower

Answer: a

38. Which of the following is not one of Sternberg and Wagner's three types of intelligence?

a. academic
b. practical
c. creative
d. musical

Answer: d

39. The Harlows' research on monkeys showed that _______ was most important in developing attachment.

a. food source
b. negative reinforcement
c. contact comfort
d. none of the above

Answer: c

40. Children in which stage of development will attain conservation?

a. formal operational
b. sensorimotor
c. preoperational
d. concrete operational

Answer: d

41. Which of the following therapeutic techniques is based in part on the concept of unconscious conflict?

a. cognitive therapy
b. psychoanalysis
c. gestalt therapy
d. b and c only

Answer: b

42. Which of the following operates on all three levels of consciousness (i.e. the conscious, unconscious, and preconscious)?
a. id
b. ego
c. superego
d. inflated ego
Answer: b

43. Tricyclics are most commonly used to treat which type of disorder?
a. depression
b. bipolar disorder
c. schizophrenia
d. obsessive-compulsive disorder
Answer: a

44. According to Selye's model (called the general adaptation syndrome), the first phase of response to stress is called the
a. alarm reaction.
b. stage of resistance.
c. stage of exhaustion.
d. none of the above
Answer: a

45. If a drug blocked reuptake of dopamine, it would be called a dopamine
a. inhibitor.
b. blocker.
c. agonist.
d. antagonist.
Answer: c

46. Cerebral specialization
a. is stronger in adults than in children.
b. is studied in split-brain patients.
c. is related to "handedness."
d. all of the above
Answer: d

47. Which people seem to be buffered against the negative effects of stress? People who are (have)

a. high self-monitors.
b. external locus of control.
c. hardy.
d. none of the above

Answer: c

48. The things that our unconscious is trying to tell us when we dream are called the _______ content of the dream.

a. latent
b. manifest
c. manifold
d. manicotti

Answer: a

49. Which of the following is not a premise of spreading activation?

a. the state premise
b. the particle premise
c. the representational premise
d. the process premise

Answer: b

50. Dorothea Dix is most well known for

a. challenging Freud's notions of psychosexual development.
b. working to get mental hospitals built in the 1800s.
c. developing logotherapy.
d. being the first person to successfully use the insanity defense in a trial.

Answer: b

51. According to Aaron Beck, which of the following is not a common irrational thought pattern of depressed individuals?

a. all-or-nothing thinking
b. overgeneralization
c. arbitrary inference
d. All of the above are common.

Answer: d

52. ________ is also know as stimulus-response (s-r) psychology.
a. Behaviorism
b. Functionalism
c. Cognitive psychology
d. Structuralism
Answer: a

53. If a split-brain patient was quickly flashed the word heart in the center of his or her field of vision, what would the person tell you he or she saw?
a. heart
b. he
c. art
d. nothing
Answer: c

54. Which of the following substances affects the norepinephrine system in the brain?
a. botulinus toxin
b. amphetamine
c. curare
d. cobra poison
Answer: b

55. Black widow spider venom causes painful convulsions by doing what to acetylcholine in the brain?
a. blocking the receptor sites
b. blocking the release
c. blocking the breakdown
d. none of the above
Answer: c

56. People suffering from Parkinson's syndrome have too little ________ in their brains.
a. serotonin
b. dopamine
c. norepinephrine
d. GABA
Answer: b

57. The phenomenon of drug tolerance (in which addicts need larger and larger doses in order to feel the same effect) is a product of

a. observational learning.
b. the dopamine system.
c. operant conditioning.
d. classical conditioning.

Answer: d

58. If told to use this strategy, young children will say a phone number over and over again to help them remember it. However, if later given a locker combination to remember, they will not automatically use the repetition strategy. This is due to a failure of

a. encoding.
b. storage.
c. retrieval.
d. meta-cognition.

Answer: d

59. The equipotentiality principle

a. states that any CS would be as good as any other to pair with the UCS.
b. contradicts the notions of preparedness and belongingness.
c. was disproven by Garcia and Koelling's study that gave rats "bright/noisy/tasty" water.
d. all of the above

Answer: d

60. A hypothesis is

a. a defense mechanism.
b. a testable prediction.
c. a conclusion based on data.
d. none of the above

Answer: b

61. An experimental method that is designed to explore cause and effect is the _______ method.

a. descriptive
b. correlational
c. experimental
d. all of the above

Answer: c

62. The pituitary is part of the _______ system.

a. limbic
b. endocrine
c. central nervous
d. sympathetic

Answer: b

63. The brain structure most closely associated with memory is:

a. the hippocampus.
b. the cerebellum.
c. the hypothalamus.
d. the medulla.

Answer: a

64. At 10 weeks after conception, the developing child is called a(n)

a. embryo.
b. fetus.
c. zygote.
d. rug rat.

Answer: a

65. The parenting style that provides children with the greatest sense of control (and is associated with a number of positive outcomes) is the _______ style.

a. authoritarian
b. permissive
c. rejecting-neglecting
d. authoritative

Answer: d

66. Which schedules of reinforcement are likely to produce the greatest rate of responding?

a. ratio schedules
b. interval schedules
c. continuous reinforcement
d. all the same

Answer: a

67. The "peg-word" system is a form of

a. mnemonic device.
b. classical conditioning.
c. gestalt therapy.
d. conservation.

Answer: a

68. Semantic memory is a type of

a. implicit memory.
b. procedural memory.
c. long-term memory.
d. short-term memory.

Answer: c

69. Someone who brags about a large number of sexual conquests may be fixated at which stage of development?

a. oral
b. anal
c. phallic
d. genital

Answer: c

70. Which of the following is not a part of the "Big 5" model of personality?

a. agreeableness
b. openness
c. emotional stability
d. sensation seeking

Answer: d

71. Hypochondriasis is a _______ disorder.

a. dissociative
b. somatoform
c. psychotic
d. mood

Answer: b

72. Which of the following could be described as an antipsychotic drug?

a. clozapine
b. chlorpromazine
c. thorazine
d. all of the above

Answer: d

73. A macrophage is a part of the _______ system.

a. central nervous
b. immune
c. endocrine
d. limbic

Answer: b

74. When faced with the stressful task of studying for this test, you put it off and tried not to think about it. This is an example of

a. expressive resistance.
b. inhibitive resistance.
c. emotion-focused coping.
d. problem-focused coping.

Answer: c

75. The fact that a previous learned association can prevent us from seeing another association that is equally predictive is called

a. blocking.
b. overshadowing.
c. generalization.
d. discrimination.

Answer: a

76. Slot machines in Las Vegas are set up on a _______ schedule of reinforcement.

a. fixed-ratio
b. fixed-interval
c. variable-ratio
d. variable-interval

Answer: c

77. A 12-year-old child with an IQ of 75 would be said to have a mental age of

a. 6.
b. 9.
c. 10.
d. 16.

Answer: b

78. The Oedipal complex occurs in the _______ stage of development.

a. oral
b. sensorimotor
c. phallic
d. genital

Answer: c

CHAPTER 20

CENTRAL MICHIGAN UNIVERSITY

PSYCHOLOGY 100: INTRODUCTION TO PSYCHOLOGY

Hajime Otani, Professor

FIVE "MIDTERM" TESTS AND ONE CUMULATIVE FINAL ARE GIVEN IN THIS COURSE. THE midterms include multiple-choice as well as short-answer sections, while the final is exclusively a multiple-choice exam. A midterm exam is included here.

The main goal of the course is to present a broad overview of psychology, introducing students to psychology as a scientific discipline. I would like students to understand how psychologists use scientific methods to investigate various phenomena. I also want them to learn various terms used in psychology.

Students need to develop a habit of studying every day. The literature is rife with studies that indicated that distributed practice is superior to massed practice. Students are much better off if they learn a small chunk at a time. Also, the same literature indicates that organization of the material is very important. I often see students being swamped with a huge amount information, which they cannot effectively process. The problem is that they are not organizing the material effectively. I often tell them that creating a story would reduce the absolute amount of information they have to remember. I also tell them that good comprehension leads to better memory. As with remembering nonsense syllables in an experiment, it is more difficult to remember material they do not understand.

I find that my students are not very good at making judgment of learning (JOL); that is, they cannot make an accurate judgment that they have learned the material. Memory researchers discovered that JOL is more accurate if it is made after a short delay. Asking yourself whether you learned the material immediately after you studied is not as accurate as asking yourself the same question after a short delay, even

20 minutes after you finish studying. So I advise my students to make JOL not while they are studying but shortly after they finish studying.

TEST 3

Short-Answer Questions

1. **According to the levels-of-processing model, some forms of processing are shallow and others are deep. What are "shallow" and "deep" levels of processing?**

Answer:

Shallow processing means you process the perceptual aspects of the material, whereas deep processing means you process the meaning of the material.

You need to state exactly what kind of information you process when you process at shallow and deep levels. To be successful in answering this question, you need to know the levels-of-processing theory. I usually conduct a demonstration of this theory. Students will be successful if they remember what we did in the classroom.

2. **According to your instructor, two types of retrieval are *explicit* and *implicit*. Describe what these are. How do we test them?**

Answer:

Explicit retrieval means you consciously recollect or retrieve past experience, whereas implicit retrieval means you do not have conscious awareness of retrieving past experience. Direct tests (recall and recognition) are often used as explicit tests, whereas indirect tests (word stem completion or word fragment completion) are often used as implicit tests.

You need to know what these retrieval types mean. Also, you have to remember what kind of tests are commonly used. Remembering the lecture about amnesia would help.

3. **Wallas suggested that creative problem solving proceeds in four steps. What are these steps?**

Answer:

Preparation, incubation, illumination, verification

Here, you need only list the names of the stages. You simply need to remember what they are.

4. **According to your instructor, Skinner and Chomsky disagreed with each other on how language is acquired. Describe their theories.**

Answer:

Skinner proposed that language is learned based on reinforcement and punishment, whereas Chomsky proposed that language acquisition is done by a language acquisition device (LAD).

You need to understand *how* Skinner and Chomsky differed in their theory regarding how language is acquired.

5. **Galton began testing people's intelligence because he wanted to implement his program of eugenics. What is eugenics?**

Answer:

Selective breeding of human beings.

Here you need to remember the lecture regarding the origin of intelligence tests. Why would anyone want to measure people's intelligence? You need to remember that Galton was serious about improving the human gene pool, but to do so, he had to measure the intelligence of those people who might be eligible for reproducing.

Multiple-Choice Questions

Tip for taking a multiple-choice test: Cover up the choices until you answer the question. The more you read the choices, the more confusing they become. Treating this test as a recall test makes it much easier.

1. **In the stage theory of memory, what is a major difference between the stages?**
   a. that control processes can only occur during the first stage
   b. the length of time information is stored during each stage
   c. that rehearsal can happen only during the first stage
   d. the accuracy of the memory trace

Answer: b

In the stage theory of memory, the major difference among stages is the amount of time information is stored in the stages.

TIP: You can eliminate a and c right away because control processes occur in all stages and rehearsal occurs only in short-term memory (the second stage).

2. **Which memory stage "briefly holds an exact image of each sensory experience until it can be processed"?**
   a. short-term memory
   b. midterm memory
   c. long-term memory
   d. sensory register

Answer: d

> **The key terms are "brief" and "exact image of each sensory experience." You have to remember that short-term memory is primarily auditory and long-term memory is primarily semantic. This leaves sensory register.**

3. **Which control process would be used to maintain a phone number in short-term memory?**
   a. rehearsal
   b. encoding
   c. decoding
   d. chunking

Answer: a

> **Rehearsal prevents forgetting in short-term memory.**

4. **As often as it is possible, we store information in short-term memory in the form of**
   a. sensory vibrations.
   b. acoustic codes.
   c. images.
   d. episodes.

Answer: b

> **Acoustic code is used in short-term memory. You can immediately eliminate a and d because these are not the codes. The choice c is more difficult to rule out, because we do use images in our short-term memory—but acoustic codes (b) are more common.**

5. **Your class was required to memorize the Preamble to the United States Constitution. Your friend tried to memorize it by reading it over and over. You, however, memorized the first sentence, then the second, then the third, and so on, until you could put it all together. You were using**
   a. reconstructive memory.
   b. a sensory register technique.
   c. transference.
   d. chunking.

Answer: d

This question is not as clear as it could be, since putting things together may imply "narrative chaining." But here the best choice is d. You can eliminate b and c quite easily. The choice a is more difficult.

6. Knowing how to play tennis is an example of _______ memory.
   a. declarative
   b. episodic
   c. semantic
   d. procedural

Answer: d

Here you need to know different types of memory. Doing things means "procedural." The only way you can answer this question is to know what kind of information is stored in each of these memories.

7. Mary, whose hobby is the study of whales, tells her teacher all about the beluga whale, even though she has never actually seen one. Mary's information is stored as _______ memory.
   a. cued
   b. motivated
   c. episodic
   d. semantic

Answer: d

Again, you need to know what kind of information is stored in each of these memories.

8. The memory of experiences that can be defined in terms of time and place is called
   a. eidetic.
   b. semantic.
   c. episodic.
   d. consolidated.

Answer: a

Factual knowledge is called for here.

9. Which memory organization theory claims that concepts are linked by experience?
   a. levels-of-processing mode
   b. spreading activation model
   c. stage theory of memory
   d. elaboration model

Answer: b

Choice b is the only model listed that proposes how concepts are linked together.

10. What is an associative network?
a. a type of recall strategy
b. a theory related to proactive interference
c. a theoretical way of organizing LTM
d. a synaptic theory of memory storage
Answer: c

You need to know what an associative network is.

11. Essay examination is to multiple-choice test as
a. recognition is to relearning.
b. recall is to relearning.
c. recognition is to recall.
d. recall is to recognition.
Answer: d

An easy question. Here you need to know different types of retrieval.

12. The tendency to recall the beginning and the end of a list of items is known as
a. the serial position effect.
b. retroactive interference.
c. motivated forgetting.
d. spreading activation effect.
Answer: a

This is an easy question. I often do a demonstration for this. To answer this question successfully, you need to know exactly what this effect is called.

13. Sometimes you cannot quite recall a word, but you remember what letter it begins with. This is called the _______ phenomenon.
A. retrograde
B. serial position
C. tip-of-the-tongue
D. semantic code
Answer: c

Everyone experiences this phenomenon. You need to know the name of it.

14. In preparing for any test, which of the following would be most beneficial to you?

a. maintenance rehearsal
b. elaboration
c. chunking
d. relearning

Answer: b

This is a difficult question. All the choices involve some ways of improving memory, but, to take a test, you need to do something to help your long-term recall. Choices a and b are the ways to improve short-term memory. Choice d is the way to improve speed of learning. Choice b is the only way to improve retrieval from long-term memory.

15. What causes forgetting to occur, according to the decay theory?

a. unpleasant experiences
b. the passage of time
c. unconscious motivations
d. interference by new material

Answer: b

You need to know what each theory proposes. The decay theory proposes that memory traces disappear as a function time.

16. The interference theory of forgetting states that

a. memory traces decay as a result of time.
b. one tends to forget unpleasant events.
c. similar memories interfere with the retrieval of one another.
d. proactive interference is created by later learning.

Answer: c

Again, you need to know what each theory of forgetting proposes.

17. Recently Tom's five-speed sports car was in the shop for needed repairs, and Tom borrowed a large car with an automatic transmission. Tom kept trying to shift gears as he would have with his stick shift, demonstrating

a. proactive interference.
b. retroactive interference.
c. retrograde amnesia.
d. episodic memory.

Answer: a

You need to know how proactive and retroactive interference work.

18. Barney was recalling the many different cars he and his family had owned over the last three decades. He found it relatively easy to describe in detail the recent cars, but it took quite a while to come up with a description of a car his parents had when he was an adolescent. His difficulty was due to
   a. anterograde amnesia.
   b. motivated forgetting.
   c. proactive interference.
   d. retroactive interference.

Answer: d

See comment on question 17.

19. According to the _______ theory, memories become _______ over time.
   a. interference; faded
   b. reconstruction; distorted
   c. decay; distorted
   d. motivated forgetting; faded

Answer: b

This is a trick question. You need to distinguish the terms *"faded"* and *"distorted." Faded* implies disappearance. *Distorted* means the memory is still there but in a different form. The trick is to eliminate choices a and c. The decay process does not "distort" memory. *Interference* does not imply *disappearance.* Also, motivated forgetting does not eliminate memory.

20. An interesting study by Marigold Linton, in which she tested her memory for events by randomly selecting 150 cards that listed her prior life events and served to jog her memory, seems to lend support to which theory of forgetting?
   a. decay theory
   b. interference theory
   c. reconstruction theory
   d. motivated forgetting theory

Answer: d

This is a question about repression. When repression occurs, you lose conscious access to memory. To gain access to repressed memory, you need to use some kind of "reminders."

21. After 15 years you can still recall in detail the moment your favorite dog was run over by the mail carrier right before your eyes. This is a(n) _______ memory.
   a. flashbulb
   b. procedural
   c. emotive
   d. repressed

Answer: a

You simply need to know the name of this phenomenon. This is an easy question, if you have studied.

22. Long-term potentiation has been demonstrated in
    a. amnesia patients.
    b. working memory tasks.
    c. hippocampal neurons.
    d. sensory receptors.

Answer: c

You need to know what long-term potentiation (LTP) is. Then you have to remember the study that demonstrated LTP. This is a relatively difficult item because the question does not provide self-evident cues.

23. Suppose a person is injected with a drug that leaves current memory intact, but the drug prevents new information from passing from short-term to long-term memory. This drug would produce
    a. anterograde amnesia.
    b. retrograde amnesia.
    c. relearning interference.
    d. proactive amnesia.

Answer: a

Every student is fascinated with amnesia. You need to distinguish between two types. One is anterograde, and the other is retrograde.

24. While delivering his papers, Ernie had a bicycle accident and received a hard blow to the head. As a result, he could not remember the events preceding the accident. Ernie is experiencing
    a. anterograde amnesia.
    b. retrograde amnesia.
    c. retroactive amnesia.
    d. episodic amnesia.

Answer: c

See the comment on question 23.

25. All people who are either big or tall, or both big and tall, is an example of a _______ concept.
    a. conditional
    b. natural
    c. conjunctive
    d. disjunctive

Answer: d

You need to know different types of concepts. It helps to convert the terms "big" and "tall" to a and b. A conjunctive concept means a *and* b. A disjunctive means a *or* b *or both*.

26. Concept formation is a special type of thinking that involves testing
    a. hypotheses.
    b. limits.
    c. validity.
    d. expert systems.

Answer: a

This is an easy question. Choice a is the only choice that is involved in concept formation.

27. Rosch suggests that some members of concepts are better examples than others. This "better example" of something in a category is termed a
    a. concept.
    b. prototype.
    c. morpheme.
    d. mental set.

Answer: b

You need to know the correct term for the "better example." It also helps to know what Rosch did.

28. A person who cannot solve a problem because he or she cannot imagine new ways of approaching a problem is exhibiting
    a. a mental set.
    b. convergent thinking.
    c. reactive fixedness.
    d. conjunctive fallacy.

Answer: a

Requires knowledge of the term. Be careful to avoid choice c, which uses the term "fixedness." Another form of mental block is called "functional fixedness."

29. You make your famous strawberry banana daiquiris for a gathering of your best friends. You use a recipe to ensure the correct combination of ingredients. Which approach to problem solving was used in this example?

a. trial and error
b. heuristics
c. expert systems
d. algorithm

Answer: d

The key here is to "ensure the correct combination of ingredients." "Ensure" automatically requires an algorithm. Of course, trial-and-error might be fun. But wrong.

30. Voting for a candidate based on having heard her on TV rather than on having systematically studied the viewpoints of all the candidates is problem solving based on

a. algorithms.
b. convergence.
c. divergence.
d. heuristics.

Answer: d

Nonsystematic ways of solving problems are called *heuristics*.

31. Which type of heuristic suggests that decisions are made based on how similar the sample is to the population from which it comes?

a. representativeness
b. availability
c. algorithm
d. trial and error

Answer: a

32. A type of heuristic that uses relevant information accessible in memory is called the _______ heuristic.

a. representativeness
b. declarative
c. availability
d. perceptual

Answer: c

33. With respect to types of thinking, rigid is to flexible as

a. divergent is to functional fixedness.
b. functional fixedness is to convergent.
c. convergent is to divergent.
d. divergent is to heuristic.

Answer: c

You need to know about two types of thinking, convergent and divergent.

34. A teacher may insist that diagramming complex sentences helps students comprehend the _______ of the English language.

a. phonology
b. morphology
c. semantics
d. syntax

Answer: d

You need to know what each choice means. Diagramming is done to understand how a complex sentence is formed.

35. The notion that the more vocabulary or categories a language has, the more varied our perceptions will be, is a theory connected to

a. Rosch.
B. Chomsky.
c. Miller.
d. Whorf.

Answer: d

You need to know who proposed the linguistic relativity theory. Eliminating choice b is difficult because Chomsky was one of the most influential language researchers.

36. The term "g" refers to the idea that intelligence

a. is made of genetically inherited abilities.
b. may be grouped into subcategories.
c. has a basic general component.
d. was developed by Sir Francis Galton.

Answer: d

The term "g" was proposed by Charles Spearman in reference to Galton.

37. According to Sternberg, encoding, inferring, mapping, applying, comparing, and responding are all
   a. aspects of long-term memory.
   b. cognitive steps in intellectual behavior.
   c. components of functional fixedness.
   d. skills used in deep structuring.

Answer: b

You need to know the theory proposed by Sternberg.

38. The ability to learn or invent new strategies for dealing with new problems is called
   a. a mental set.
   b. expertise.
   c. fluid intelligence.
   d. specific intelligence.

Answer: c

You need to know that there are two types of intelligence, called *"fluid"* and *"crystallized."* Fluid intelligence is needed to deal with new concepts.

39. If a child's mental age were higher than his or her chronological age, this would mean that
   a. the child is brighter than normal.
   b. the child is about average.
   c. the child is less intelligent than normal.
   d. a calculation error was made.

Answer: a

Requires factual knowledge of how IQ is calculated.

40. Reading the identical directions to every person taking a given intelligence test is an aspect of the test's
   a. reliability.
   b. validity.
   c. normalization.
   d. standardization.

Answer: d

This question is about what makes a test a good test.

41. When a test is given to a large group of individuals varying in age, sex, background, and so on, to form a basis for interpreting an individual's score, the group is referred to as the ________ sample.
    a. standardization
    b. validity
    c. affirmative
    d. normative

Answer: d

Another question on the elements of creating a good test.

42. If a person receives similar scores on a test that she takes over and over on several different days, the test is said to be
    a. objective.
    b. valid.
    c. reliable.
    d. normed.

Answer: c

A measure you use must be reliable. If a ruler is not reliable, you never know how long a stick is.

43. Suppose you were given a comprehension intelligence test, but you realized while taking the test that it was not measuring intelligence, just reading speed. Afterward, you did not feel that the test measured what it was intended to measure. In other words, you felt that the test was not
    a. standardized.
    b. valid.
    c. reliable.
    d. normative.

Answer: b

Here's another important factor in constructing a test: a test must be valid in order to measure what you want to measure.

44. Correlations between the intelligence scores of identical twins tend to be
    a. high.
    b. low.
    c. the same.
    d. normative.

Answer: a

You need to remember the studies (discussed in class and in the textbook) that compared twins.

45. Adoption studies have suggested that

a. heredity plays a slight role in intelligence.
b. heredity plays an important role in intelligence.
c. environment plays no role in intelligence.
d. environment and heredity play equal roles.

Answer: b

This is a tricky question. You need to know why anyone would conduct an adoption study and also the general findings of such studies.

46. Which of the following is true concerning cultural differences in intelligence scores?

a. Available data support a genetic interpretation.
b. The gap between white and black scores has widened.
c. Available data support an environmental interpretation.
d. The differences are intellectually based, not cultural.

Answer: c

Again, you need to know the general findings of the studies that examined cultural differences. This is a relatively difficult question.

47. Of the four designations of retardation, which accounts for the greatest percentage of individuals who are considered retarded?

a. profoundly
b. severely
c. moderately
d. mildly

Answer: d

Requires knowing that the milder form is more common than the severe form. Also requires knowledge of the different degrees of retardation.

CHAPTER 21

CENTRAL MICHIGAN UNIVERSITY

PSYCHOLOGY 100: INTRODUCTION TO PSYCHOLOGY

Debra Ann Poole, Professor

MY PRIMARY OBJECTIVES ARE

- To give students an understanding that psychology is a behavioral science that encompasses a broad range of topics (i.e., psychology is not just clinical psychology)
- To develop an appreciation for the empirical study of human behavior and the importance of critical thinking
- To develop a set of basic knowledge about psychology that will support future learning (i.e., to transmit a core set of fundamental facts that will allow them to explore psychological findings after they leave the class)
- To build a stronger English vocabulary that will support learning across disciplines.

Students should learn that human behavior is guided by principles that can be studied empirically, just as we study the physical world. As this understanding develops, they should begin to distinguish between information that results from a scientific process and pop psychology. They learn that intuition is not a reliable source of information about human behavior. They also learn that individual studies are limited and can lead to incorrect or incomplete conclusions, but that the scientific process prevails over intuition because it progresses toward greater understanding when converging evidence is the basis for decisions. Within this overarching framework, students need to develop a working knowledge of the vocabulary and core phenomena in psychology, focusing in foundation courses on information that is useful for being an informed consumer of scientific knowledge across disciplines. Because students often begin college without an appreciation of how information is organized into concepts, the presentation and testing of material should lead them to see information as organized into general principles and supporting observations, and they should be able to discuss the relationships between concepts and specific observations.

Students often come to college with very little experience learning on their own. They are accustomed to teachers who explicitly transmit every key term, and many students learn from texts predominantly by memorizing facts as unrelated pieces of information. In college, however, the quantity of material is too great to continue this strategy. Students need to learn to monitor their own understanding of material by engaging in a dialogue with themselves about how information is organized into concepts, and how those concepts relate to what they already know. The most useful first step is to learn to approach a text in a new way. I teach them to scan the chapter outline, then turn immediately to the summary, read it, and turn to another subject for that evening. This gives them an overview of what the major concepts in the chapter are and allows them time to think about those concepts without interference from reading related information. I ask them next to take several evenings to read the chapter, taking care to read the comics, illustrations, and special topic boxes that help them understand and remember the information. After reading the summary again, they need to master the critical terms by engaging in an internal dialogue about those terms. They might, for example, write and answer their own essay question about classical conditioning that requires them to label the major terms involved in that form of learning, or they might rehearse the neuron by drawing and labeling a diagram. They need to realize that study guides are generally ineffective for initial learning of material, but that they are very effective to help students identify gaps in their knowledge after they believe that they have mastered the material.

ESSAY QUESTIONS

Thirteen weekly quizzes are given in this course, plus an optional comprehensive final, which can count for up to two missed exams. Each 20-point examination consists of fourteen multiple-choice questions and two 3-point essays. The multiple-choice questions consist of "fact" questions, which require students to master basic vocabulary, and "concept" questions in which students must recognize the concept that underlies a term when that concept is represented by an example or is described in a new way. The essay questions encourage students to organize information (predominantly from lectures) into concepts. They might be required, for example, to compare and contrast two terms or approaches to an issue, to describe research that supports a counterintuitive conclusion, or to apply information learned in class to a new situation. Examples of two essay questions, each with two alternative responses, are presented here.

Essay Question 1

Contrast at least three differences (each contrast on a concept is one difference) between schizophrenia and multiple personality disorder. Write in essay format.

This is a compare/contrast question, and the best strategy is to choose one concept (e.g., age of onset), contrast each disorder on that concept, and proceed to the next concept. When a student simply describes schizophrenia and then describes multiple personality disorder, it is easy to err by not including parallel information for the two disorders.

A successful answer must provide three valid contrasts between schizophrenia and multiple personality disorder, with each contrast focusing on a single concept (e.g., symptoms, "age of onset," sex ratio). An excellent answer would go beyond a simple sentence about each of these and would explain caveats. For example, there is debate about the existence of MPD as a disorder that stems from trauma, and therefore the assumption that it begins in childhood (although often not diagnosed until later) rests on the trauma argument.

Answer A:

Schizophrenia is a cognitive disorder where hallucinations and delusions rule their reality. The disorder affects those in late adolescence or early adulthood and is believed to be brought on biologically. Treatment is drugs. Multiple personality disorder is a dissociative disorder with cognitive thinking intact. It is brought on in early childhood (majority women) brought on by a traumatic (environmental) experience. Its treatment is talk therapy.

This student mentioned three valid contrasts and received full credit, but it is far from a stellar answer. The reader had to match up the concepts across diagnostic categories, whereas it would have been better to mention a single concept, such as "age of onset," and contrast it across the two diagnostic categories. The student also gave sweeping generalizations, such as that schizophrenia is always associated with hallucinations and delusions. Moreover, the student does not fully define terms. "Dissociative disorder with cognitive thinking intact" is ambiguous. What is a dissociative disorder? What is meant by "thinking intact." If this question appeared as a major essay or in an upper-level class, I would have taken points off for failing to develop a complete answer.

Answer B:

Multiple personality disorder (MPD) and schizophrenia are two disorders that are near the opposite ends of the mental illness spectrum. Regardless of this fact, they sometimes still are misused terms. There are some significant differences between the two. First, schizophrenia's sex ratio is about 50% male and 50% female. However, MPD is predominantly female. Secondly, to treat these disorders, schizophrenics are usually treated medically with drugs. MPD is usually treated with psychotherapy or talk therapy. Finally, the age of onset of most schizophrenics is during adolescence. However, most MPD victims show signs during childhood.

This illustrates an improved essay. The student begins with a topic sentence and tackles a single concept at a time. She also alludes to the controversy about MPD by placing the word "victims" in quotation marks.

This essay could have been improved by more discussion. For example, the difference in sex ratio for cases diagnosed might reflect who seeks treatment rather than true differences in the underlying populations, and the student could have discussed the assumption that MPD arises in childhood by describing the alternative perspective that some cases arise due to social processes in therapy.

Essay Question 2

Explain what the "diathesis-stress" model of mental illness is (i.e., vulnerability-stress) and give an example of research data that supports this model. (Hint: the twin research presented on Wednesday, the depressed monkey study, etc.)

It is important first to lay out the basics by defining what each component of the model refers to, and by giving an example. The supporting research must be described accurately, but in concluding the essay, the student must relate that research back to the two central concepts. Many students lost points on this essay because they simply described a study, but did explain how that study demonstrated both "diathesis" and "stress." These students often remembered an example from class, but had no idea what the concept from that class period really was. That is, they did not comprehend the point that many psychological and medical disorders arise from the joint contribution of several factors. Students need to ask explicit questions in class, such as, "What is the purpose of this example?"

A successful answer would begin by explaining each component term in the model; that is, that diathesis refers to a biological vulnerability and that stress refers to some trigger in the environment that initiates the disorder. A good answer would go beyond this and give examples of each of these. For example, a biological predisposition could be vulnerability to a specific virus (due to an inherited characteristic of the immune system) or an abnormal level of a specific neurotransmitter in the brain. An environmental trigger might be exposure to a specific virus, or social stress that influences neurotransmitter levels. The successful answer must then explain one set of research findings that supports that model, and why. I prompted students to consider using one of two examples we discussed in class: (1) concordance rates for monozygotic and dizygotic twins supports the model in the case of schizophrenia, and (2) animal analog studies have simulated a biological predisposition by drugging one group of monkeys to study depression, but the difference between groups is not detectable unless a stress (separation from mom) is introduced.

Answer A:

The diathesis-stress model of mental illness uses two factors. This model shows that (1) a person is medically/biologically predisposed for the disorder and (2) has the stress or an environmental trigger. Diathesis is the biological predisposition and stress refers to the environmental trigger. Research data that supports this is the twin research. Looking at monozygotic (MZ) and dizygotic (DZ) twins, the concordance between MZ both having schizophrenia is higher than DZ's concordance of having schizophrenia. This shows that it must be partially due to genetics or biological predisposition. However, there might just be a stress such as an environmental trigger. The data supports this because it is not 100% biological with MZ twins.

This student received full credit. She started by explaining each term in the model and linked the twin methodology to that model. Her answer could have been improved, however, by further discussion. I would like to have seen examples of "diathesis" and "stress" from class discussion before jumping into the twin methodology. Also, I would like to have seen more discussion about the assumptions that underlie the twin methodology, such as what "concordance" rates are and why comparing across MZ and DZ twins informs us about etiology (i.e., the causes of a disorder).

Answer B:

The diathesis-stress model of mental illness is divided into two parts: a biological factor and a trigger factor. The point of the model is to show that due to a trigger in predisposing (biological) factors depression can remain constant or variable. For example, rhesus monkeys were taken in a lab environment and some were given drugs (biological factor). At the time this occurred their behaviors were consistent in monkeys with drugs and monkeys without drugs. Then their mothers were taken away (trigger) and all monkeys became depressed. However, the monkeys that weren't given any drugs were able to recover to normal. But the monkeys that were given drugs were predisposed to depression and the trigger (mother taken away) kept them at the depressed state and they never recovered. Therefore, predisposing factors may inhibit long-term depression in people.

This student understands the central concepts and received full credit, but failed to proofread the essay. The second sentence of the essay makes little sense, and the final sentence is confusing. It appears that this student wrote from head to paper without stopping to think about how else he might have worded each sentence. For example, "their behaviors were consistent in monkeys with and without drugs" means that before stress was introduced, the behavior of the two groups of monkeys was identical, indicating that the biological predisposition alone was insufficient to cause a difference in behavior. This student could write clearer essays by first outlining his thoughts, then stopping to think about alternative ways to phrase each sentence.

When approaching essay questions, students should first read the question and ask, "What is this question asking me to do?" They should then take a minute or two to plan or outline their answer by making notes on their test. Students often lose points for mentioning a term without demonstrating that they know what it means. Each term mentioned should be clearly defined and, if possible, illustrated with an example. Students should reread their answers, trying to take the perspective of a student who has not taken the class. In other words, they should not assume that the faculty member will make connections for them. If their roommate would not understand the point of their essay, then the faculty member cannot give them credit for understanding it.

The biggest problem that students have with essay questions is that they do not put down what they know, but, rather, they assume that the reader will fill in the gaps. They often answer with sketchy comments and fail to define each term that they mention. They forget that the overall goal is to prove to the faculty member that they know the material.

Students often see a short essay as a "short answer" and fail to develop a logical argument. Students should realize that class members who are faking it often mention terms that appeared in the multiple-choice questions, hoping to get a point for stating a critical term. To receive credit, though, a student needs to demonstrate that he knows what that term is and how it relates to the question that was asked. I grade leniently for a freshman class in the fall semester, but for more advanced classes the overall quality of the essay will factor into their grade.

CHAPTER 22

DEPAUL UNIVERSITY

PSYCHOLOGY 106: INTRODUCTORY PSYCHOLOGY II

Fred Heilizer, Associate Professor

FROM THE COURSE SYLLABUS:

Either Psychology 105 or 106 is required for any advanced psychology course. If you *are* a Psychology major, *both* 105 and 106 are required for graduation. If you are not a Psychology major, *either* 105 or 106 may be used to satisfy [the core] requirement.

The Introductory Psychology course is intended to present (a) basic psychological methodology and terminology, (b) a broad variety of psychological knowledge and theory, and (c) integration of various approaches and areas of psychology. As a result of having taken—and passed—this course, you can expect to have a pretty good knowledge of basic psychology and its many, many, many ramifications.

You should complete the assigned readings and take a midterm and final exam based on the readings and on class presentations. It is generally suggested that two hours of reading and study be done for each class hour . . . using an active process such as SQ4R. The midterm and final exams are mostly objective, multiple-choice tests, but they also may include one or more essay questions. Each test covers half the course material with each counting equally to the final grade. The test grades will be largely curved to the class's test performance.

The "SQ4R" mentioned in the syllabus stands for Survey, Question, Read, Recite, Write, Review. It involves several passes through a chapter, with active participation in the process. Another useful study practice is called "Never Just Read a Chapter," which emphasizes starting with learning the summary or summaries of a chapter and organizing the major and minor headings of the chapter in terms of the content and organization of the summary while actively making notes in the margins or separately.

As for class attendance and lecture notes, I suggest that the class coverage represents my effort to organize and extend the text's coverage. Therefore, good class notes represent the best study outline available to the student. In addition, I strongly urge students to develop their own version of speedwriting for use in this class and others, and I illustrate a speedwriting technique throughout the quarter.

Two exams are given in the course, a midterm and a final. Together, the exams determine the bulk of the course grade.

MIDTERM EXAM

1. Sleep apnea tends to occur with

a. overweight males.
b. unmarried women.
c. frightened children.
d. overworked college students.

Answer: a

2. A basic law of neural transmission is

a. more-or-less.
b. all-or-none.
c. before-or-after.
d. early-or-late.

Answer: b

It is important to understand the binary (all-or-none) nature of neural transmission at the level of the individual neuron.

3. Drug addicts are especially sensitive to

a. REM activity.
b. constancy effects.
c. triggers.
d. hypnotism.

Answer: c

4. Endorphins are

a. analgesic drugs such as morphine and opium.
b. endocrine glands.
c. metabolizers of alcohol.
d. neurotransmitters that act similarly to opiates.

Answer: d

5. **An individual nerve cell is called**
   a. a cell body.
   b. an axon.
   c. a neuron.
   d. a dendrite.

Answer: c

6. **Early sleep is primarily characterized by the occurrence of**
   a. stage 4.
   b. REM.
   c. nightmares.
   d. dreaming.

Answer: a

This can be tricky if you don't understand that stage 4 sleep is reached fairly rapidly.

7. **The onset of alcoholism is often characterized by**
   a. the sacramental use of wine.
   b. drinking with meals.
   c. binge drinking.
   d. drinking beer at ball games.

Answer: c

When all else fails, look for the most sensible answer.

8. **The major connection between the two hemispheres of the cerebral cortex is the**
   a. corpus callosum.
   b. cerebellum.
   c. sensory viaduct.
   d. occipital lobe.

Answer: a

9. **One interpretation of hypnosis, coming from Orne's research, is that hypnotized people are**
   a. schizophrenic.
   b. playing a role.
   c. weak minded.
   d. neurotic.

Answer: b

10. In order for a drug addict to finally kick the habit, he or she must develop a sense of
- a. imminent death.
- b. danger.
- c. financial stress.
- d. being accepted.

Answer: d

11. Senden's review of clinical research with newly sighted cataract patients indicated that we have to learn to perceive
- a. colors.
- b. shapes and constancies.
- c. motion.
- d. figure and ground.

Answer: b

12. A sleep cycle lasts for about
- a. 30 minutes.
- b. 90 minutes.
- c. 2 hours.
- d. half the night.

Answer: b

13. If we traveled to a planet in a different solar system, which of these would be most true? We would see
- a. no new colors.
- b. many new colors.
- c. only new colors.
- d. no colors at all.

Answer: a

The answer is based on the universal nature of electromagnetic radiation, including visible light, and on the fact that a trip to another planet would not change the organs with which we perceive light (and, therefore, color).

14. Experimental method is defined as the systematic use of
- a. correlations.
- b. independent variables.
- c. measurements.
- d. dependent variables.

Answer: b

15. A basic law of the nervous system is one-way transmission. This is due to the way transmission occurs at the

a. synapse.
b. interneuron.
c. myelin sheath.
d. cell body.

Answer: a

All you need to know in order to answer this item is the site of neural transmission.

16. The major operation, or variable, in perception involves the manipulation of

a. memory.
b. the sensory homunculus.
c. reinforcement.
d. attention.

Answer: d

17. Cones are

a. sensitive to dim light.
b. more numerous than rods.
c. responsible for color vision.
d. very important for audition.

Answer: c

18. Aside from normal-type dreams, most sleep functions occur in:

a. Stage 1, REM.
b. Stage 1, non-REM.
c. Stage 2.
d. Stage 4.

Answer: d

19. In most adult humans, the left cerebral hemisphere is dominant in the function of

a. intuition.
b. speech.
c. audition.
d. emotion.

Answer: b

20. Visual acuity is highest for images projected on the

a. periphery of the retina.
b. optic nerve.
c. pupil.
d. fovea.

Answer: d

21. A neuron at rest is said to be

a. disorganized.
b. polarized.
c. defective.
d. lazy.

Answer: b

22. Genes located on the X chromosome are often causes of effects that are called

a. sex-linked.
b. chromosomal.
c. homozygous.
d. monozygotic.

Answer: a

23. Cataracts that are easily visible to another person are in which part of the eye?

a. retina
b. iris
c. cornea
d. optic disc

Answer: c

Read the question carefully. The key phrase is "easily visible." Of the eye structures listed here, the only one easily visible to another person is the cornea (c), the eye's clear outer covering.

24. One of the ways in which experiments can go wrong is through the occurrence of

a. demand characteristics.
b. double-blind methods.
c. random sampling.
d. dependent variables.

Answer: a

Note that b and c are purposeful experimental methods and that dependent variables are part of any experiment. This leaves a as the only possible answer.

25. The unfortunate historical aspect of genetics involves the social movement of

a. nurture.
b. eugenics.
c. anarchy.
d. Marxism.

Answer: b

26. The operation that severs the connections between the prefrontal lobe and the rest of the brain is called

a. genectomy.
b. lobotomy.
c. frontotomy.
d. prefrontectomy.

Answer: b

Also sometimes called *prefrontal lobotomy.*

27. Experimental method requires the use of what kind of group or treatment?

a. clinical or counseling
b. observational
c. control or comparison
d. imaginary

Answer: c

A cornerstone of the experimental method is the inclusion of a control group or control condition against which the experimental group can be compared.

28. A basic organization of nervous system structure and transmission is

a. the crossover effect.
b. reciprocal inhibition.
c. sideways transmission.
d. parallelism.

Answer: a

29. REM sleep generally occurs when the person is

a. sleepwalking.
b. dreaming.
c. hypnotized.
d. insomniac.

Answer: d

30. The fundamental Gestalt law of perception involves

a. distance and motion.
b. rods and cones.
c. four primary colors.
d. figure and ground.

Answer: d

31. Ethical considerations in human research generally require

a. psychology students.
b. informed consent.
c. payment of subjects.
d. the use of statistics.

Answer: b

The key word is *ethical*. Focus on that and then identify the only term among the choices that relates to ethics.

32. The blind spot is

a. a result of retinal damage.
b. part of the fovea.
c. in the cornea.
d. insensitive to light.

Answer: d

Often, the most obvious answer is indeed the correct one.

33. The actual appearance of an organism, reflecting the effect of genetic action, is called the

a. sociobiology of behavior.
b. proximate cause.
c. phenotype.
d. chromosome.

Answer: c

Phenotype is visible, whereas genotype is the biochemical makeup of the genes themselves, the effects of which may or may not be visibly apparent in the organism.

34. Hypnosis can be widely used to

a. give people special powers.
b. increase IQ.
c. relive early experiences.
d. reduce the experience of pain.

Answer: d

35. Hering's opponent process theory of color vision is consistent with the occurrence of

a. color blindness.
b. the action potential of the axon.
c. three types of rods.
d. bilaterality of the cortex.

Answer: a

36. Night vision is a function of

a. inhibition.
b. rods.
c. cones.
d. bipolar cells.

Answer: b

37. If we are in a situation where there are no distance cues, we would most likely

a. experience size constancy.
b. reverse figure and ground.
c. squint a lot.
d. get lost.

Answer: d

38. One of the most powerful methods of increasing healthy brain functioning, including intelligence, is

a. more sleep for infants.
b. relaxation training.
c. early perceptual enrichment.
d. hypnotism.

Answer: c

39. Night terrors generally occur in what part of sleep?

a. deep
b. light
c. late
d. REM

Answer: a

40. Experimental method requires the assignment of subjects to two or more conditions by a process that is

a. biased.
b. behavioral.
c. creative.
d. random.

Answer: d

The object of the experimental method is to avoid biases.

41. The rods and cones face in what direction with respect to incoming light?

a. upward
b. backward
c. frontward
d. downward

Answer: b

42. Perception is generally more accurate than sensation. This is best illustrated by the

a. constancy effects.
b. adaptation curve.
c. reward effect.
d. S-R model.

Answer: a

43. Accidental changes in genetic structure are called

a. artificial selection.
b. heterozygous.
c. mutations.
d. ethology.

Answer: c

44. A consistent behavior of drug addicts is
a. the use of denial.
b. a lack of dreams.
c. a high degree of hypnotizability.
d. a highly competitive attitude.
Answer: a

45. A widely used characteristic of hypnosis is
a. weak-minded hypnotic subjects.
b. the curing of addictions.
c. production of post-hypnotic effects.
d. the production of sleep without dreaming.
Answer: c

46. Problems with usage first show up as drug
a. dreams.
b. withdrawal.
c. dependence.
d. abuse.
Answer: d

47. Which of these is an important indicator of drug addiction?
a. nightmares
b. baldness
c. tolerance
d. stomach ulcers
Answer: c

48. PKU can be treated and cured by
a. a rehabilitation program.
b. diet.
c. surgery.
d. gene splicing.
Answer: b

49. The white matter of the spinal cord is where
a. fast long-range transmission occurs.
b. the sympathetic nervous system acts on the inner organs.
c. reflexes are mediated.
d. endocrine activity occurs.
Answer: a

50. The parents of a blue-eyed child
a. must include at least one with the gene for blue eyes.
b. must both have blue eyes.
c. may both have brown eyes.
d. must include at least one with blue eyes.

Answer: c

Requires understanding dominance and recessiveness, as well as the concept of phenotype versus genotype.

51. The part of the nervous system that controls the internal organs is called
a. internal.
b. autonomic.
c. endocrine.
d. automatic.

Answer: b

52. The most important first step toward being hypnotized is
a. being willing.
b. falling asleep.
c. being weak willed.
d. feeling anxiety.

Answer: a

FINAL EXAM

1. Probably the most widely used *behavioral* therapy is
a. systematic desensitization.
b. psychoanalysis.
c. client-centered counseling.
d. hypnotic age regression.

Answer: a

2. Fossil reconstructions suggest that which of the following was an essential precursor to the development of human speech?
a. an opposable thumb
b. lowering of the larynx
c. something to talk about
d. family living

Answer: b

Find the answer within the question. Fossil reconstructions could tell us nothing about c or d, and it is far more likely that b would be related to speech than a.

3. A patient who proclaims he is Jesus of Nazareth is expressing a
   a. bipolar disorder.
   b. delusion of reference.
   c. delusion of grandeur.
   d. conversion reaction.

Answer: c

4. If food-relevant cues are not present, overweight people, relative to average-weight people, will
   a. work harder for food.
   b. eat more food.
   c. feel frustrated.
   d. eat less food.

Answer: d

5. The basic n Ach law (with risk taking or goal setting) is
   a. negative.
   b. positive.
   c. curvilinear.
   d. nonexistent.

Answer: c

6. American sign language is a bona fide language that is processed in which part of the brain?
   a. left hemisphere
   b. right hemisphere
   c. premotor subcortex
   d. frontal lobe

Answer: a

A tricky question. Language is normally processed in the brain's left hemisphere, but, because sign language depends solely on visual cues, some students might assume incorrectly that it would be processed in the *right* hemisphere. Reflect on the question, which emphasizes that sign language is a "bona fide language," and you should be guided to the correct choice—provided that you know where language is processed in the first place.

7. **Minor tranquilizers, such as Librium and Valium, are also termed what kind of drugs?**
   a. antipsychotic
   b. antidepressant
   c. antianxiety
   d. endorphin inhibitor

Answer: c

8. **Feelings of being forced to repeat certain actions many times or of persistent unwanted thoughts correspond to**
   a. obsessive-compulsive disorders.
   b. phobias.
   c. panic reactions.
   d. conversion reactions.

Answer: a

9. **Dissociative disorders involve**
   a. obsessions.
   b. sudden changes in one's identity.
   c. mostly hallucinations.
   d. sexual dysfunctions and perversions.

Answer: b

10. **Seligman's model of learned helplessness has been proposed as an explanation of**
    a. dissociative states.
    b. personality disorders.
    c. schizophrenia.
    d. depression.

Answer: d

11. **Homosexuals who accept their homosexuality are**
    a. going back into the "closet."
    b. as well-adjusted as heterosexuals.
    c. less well-adjusted than heterosexuals.
    d. mostly bisexual.

Answer: b

12. Jet lag is partly caused by a disruption of

a. the opponent process.
b. mealtimes.
c. alpha rhythms.
d. enzyme secretions.

Answer: b

13. The second stage of Selye's General Adaptation Syndrome (GAS) is the stage of

a. resistance.
b. alarm.
c. exhaustion.
d. stress.

Answer: a

14. When a person is extremely manic, it is often easy to confuse the diagnosis with

a. psychogenic fugue.
b. schizophrenia.
c. obsessive-compulsive disorder.
d. personality disorder.

Answer: b

15. The capacity to form an unlimited set of sentences to express any idea is a feature of

a. all human languages.
b. chimp speech.
c. language in industrialized societies.
d. speech, but not necessarily language.

Answer: a

16. Any condition that energizes and directs an organism's actions is called a(n)

a. stress.
b. cognition.
c. instinct.
d. motivation.

Answer: d

17. An eating disorder characterized by eating binges followed by purging is

a. anorexia.
b. food phobia.
c. bulimia.
d. compulsive anorexia.

Answer: c

18. Saying whatever comes to mind without censorship or direction is

a. immature.
b. transference.
c. extroverted.
d. free association.

Answer: d

19. Primary intervention or treatment corresponds to

a. early intervention.
b. prevention.
c. psychotherapy.
d. drug therapy.

Answer: b

20. The treatment of choice for major depression is

a. systematic desensitization.
b. frontal lobotomy.
c. psychoanalysis.
d. electroconvulsive therapy (ECT).

Answer: d

21. The helping profession that is trained to prescribe drugs is

a. psychotherapy.
b. psychiatry.
c. therapy.
d. psychology.

Answer: b

22. In revising previous classifications, and greatly enlarging the coverage, the authors of DSM-III (and -IV)

a. emphasized behavioral descriptions and reliability of diagnosis.
b. eliminated the personality disorders.
c. emphasized psychoanalytic theory.
d. chose one major theory that best explained each of the important conditions.

Answer: a

The only answer that is compatible with the idea of *enlarging* coverage is a.

23. One of the moderators of (i.e., resistors to) the negative effects of stress is
a. a lot of money.
b. belief in luck.
c. sense of internal control.
d. competitive, aggressive personality.
Answer: c

24. One's sexual orientation is thought to develop in
a. infancy or early childhood.
b. puberty.
c. the latency years.
d. adulthood.
Answer: a

25. Multiple personality disorder people are likely to have had which early experience?
a. sibling rivalry
b. severe childhood abuse
c. theatrical training
d. fetal stress
Answer: b

26. Opponent process theory suggests that substances that produce intense aftereffects are
a. easily avoided.
b. potentially addictive.
c. poisonous.
d. enzymatic in origin.
Answer: b

27. Probably the most effective treatment for bipolar affective disorder (manic-depression) is
a. amphetamine or other stimulants.
b. sedatives.
c. lithium.
d. ECT.
Answer: c

28. The use of Prozac to treat depression corresponds to an intervention that is

a. tertiary.
b. motivational.
c. primary.
d. ineffective.

Answer: a

29. Psychologically disturbed people who improve without therapy are said to experience

a. extinction.
b. spontaneous remission.
c. malingering.
d. apparent retromission.

Answer: b

30. Loss of function, such as paralysis or blindness, with no physical damage is called

a. hypochondriasis.
b. malingering.
c. panic disorder.
d. conversion disorder.

Answer: d

Don't rush. Look for the best—the most precise—answer. Both a and b may seem plausible, but only d is correct.

31. The behavior therapy in which a person watches someone doing something that the person fears is

a. tokenism.
b. modeling.
c. implosion.
d. rational therapy.

Answer: b

32. The surgical procedure that was once thought to cure schizophrenia is

a. decortication.
b. EST/ECT.
c. lateral decortication.
d. frontal lobotomy.

Answer: d

33. Iatrogenic disorders are produced by

a. sex-linked genetic effects.
b. stressful environments.
c. therapies.
d. specific iatrogenes.

Answer: c

***Iatros* is the Greek word for physician.**

34. Schachter's research on obesity indicates that overweight people are likely to lack *what* with respect to their eating?

a. sensitivity to clock time for eating
b. an external orientation
c. a taste for good food
d. homeostatic controls

Answer: d

35. The psychotherapy that is most likely to directly challenge a client's irrational statements is

a. rational-emotive.
b. client-centered.
c. group.
d. individual.

Answer: a

36. A patient who sits immobile and unresponsive for long periods of time with occasional bursts of purposeless motor activity expresses which of the schizophrenias?

a. undifferentiated
b. catatonic
c. paranoid
d. hallucinatory

Answer: b

37. Multiple personality disorder people are limited to how many personalities?

a. three
b. four
c. twelve
d. an unlimited number

Answer: d

38. **Damage to the ventromedial hypothalamus (VMH) will usually cause an animal to**
    a. starve.
    b. behave aggressively.
    c. overeat.
    d. lose interest in sex.

Answer: c

39. **The sexual experience of females, as compared to that of males, is much more**
    a. varied.
    b. maladjusted.
    c. inhibited.
    d. rapid.

Answer: a

40. **Joe is more motivated by his fear of failing than by a desire to do as well as he can do. If given a choice during a game of horseshoes, he would probably stand at what distance(s) from the pole?**
    a. at the same distance as everyone else
    b. either close or at a moderate distance
    c. either far or at a moderate distance
    d. either close or far away

Answer: d

At first glance, response d would seem to defy common sense; but note that d represents two responses to fear of failure. Moving in close makes the task easier, thereby reducing the chance of failure, and moving far away makes the prospect of failure less threatening. Who will blame Joe if he misses from so great a distance?

41. **Tourette's disorder is characterized by the repeated occurrence of**
    a. delusions.
    b. obscenities and movements.
    c. hallucinations.
    d. obsessions and compulsions.

Answer: b

42. **In psychoanalytic therapy a love-hate relationship develops between the patient and his or her therapist. This is called**
    a. transference.
    b. catharsis.
    c. behavior mod.
    d. nothing.

Answer: a

43. Rapid flight of ideas and pressure of speech are symptoms of which disorder?
a. conversion
b. major depression
c. bipolar
d. multiple personality
Answer: c

44. A nonthreatening atmosphere in which people can become more aware and accepting of themselves is a central goal of which therapy?
a. client-centered counseling
b. rational-emotive therapy
c. psychoanalysis
d. behavior modification
Answer: a

45. A literature survey of 475 studies reported that the average person in psychotherapy shows more improvement than does what percent of subjects randomly assigned to no-therapy control groups?
a. 50 percent
b. 0 percent to 40 percent
c. 80 percent
d. 40 percent to 60 percent
Answer: c

46. In order for the AIDS virus to spread from one person to another, it must enter the second person's
a. hypothalamus.
b. genitalia.
c. saliva.
d. blood.
Answer: d

47. A pattern of repeated irresponsible behavior, lack of genuine relationships, impulsiveness, and freedom from guilt is which personality disorder?
a. hysterical
b. passive-aggressive
c. antisocial
d. autistic
Answer: c

48. A basic question that chimpanzee research has not conclusively answered is whether chimps can

a. talk.
b. learn language.
c. communicate.
d. use symbols.

Answer: b

Ask yourself why b is a better answer than the others, which, after all, are closely related to it.

49. Children whose parents expect them to show independent behavior at an early age are likely to develop a

a. strong achievement need.
b. low level of arousal.
c. strong fear of failure.
d. sense of rejection.

Answer: a

50. Virtually all stress-management programs include some form of

a. biofeedback.
b. hypnotism.
c. relaxation training.
d. meditation.

Answer: c

51. Solomon and Corbit's opponent process theory proposes that, with repeated exposure to a situation that produces an emotion, the initial emotional reaction will ______ while the opponent reaction will ______.

a. remain constant; grow stronger
b. weaken; strengthen
c. weaken; stay constant
d. strengthen; weaken

Answer: b

52. The sexual experience of males includes an effect that the female sexual experience does not include. This is the

a. refractory period.
b. multiple orgasm.
c. stage of excitement.
d. resolution.

Answer: a

CHAPTER 23

KANSAS STATE UNIVERSITY

PSYCHOLOGY 110: GENERAL PSYCHOLOGY

Jeffrey J. Sable, Graduate Teaching Assistant

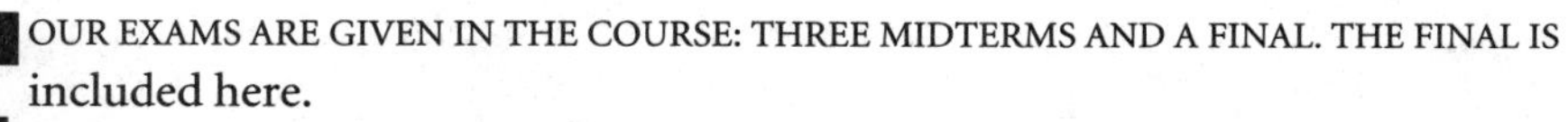
FOUR EXAMS ARE GIVEN IN THE COURSE: THREE MIDTERMS AND A FINAL. THE FINAL IS included here.

Course objectives include:

- Helping students to become familiar with the basic history, theories, and procedures of psychology
- Giving students the ability to generalize and apply psychological concepts to the "real world"
- Encouraging students to think critically about the practice and implications of psychology
- Helping students to articulate their ideas about psychology

I hope that the students will be able to see how psychology is relevant in some way to just about every aspect of their lives, and that they will be able to apply the concepts they learn in class to their own lives. I also believe it is very important for students to think and respond critically to what they see and hear about psychology. I want them to question the things they see and hear and not just blindly accept them.

The course contains a great deal of varied material. If you have any question whatsoever, ask it. Many introductory psychology courses have students from very diverse academic perspectives. It is very difficult to present effectively such a diverse subject as psychology to such a diverse group of students without some feedback. I have always found that productive questions can really facilitate understanding. Often, if one person has a question, several other students have the same question.

Try to see how each subject relates to something that interests you. Not only will you learn and remember more, but the information will also be more useful to you.

FINAL EXAM

In my course, it is important to study what was discussed in class. When writing the exams, I generally try to address concepts that were specifically addressed in the classroom. I may have two or three multiple-choice questions from the book that we did not directly address in class, just to add a little challenge and see who is really working hard.

Multiple Choice

Choose the *best* answer to each of the following and write the corresponding letter on the answer sheet (1 point each).

1. **Compared to people with an external locus of control, people with an internal locus of control tend to**
   a. develop fewer illnesses.
   b. believe that fate plays a bigger role in their lives.
   c. think the world is a hostile place that one must learn to escape from internally.
   d. be type A people.

Answer: a

2. **In general, which of the following appears to be most important in the effectiveness of psychotherapy?**
   a. a good client-therapist relationship
   b. the length of the psychotherapy
   c. the client's unconscious motives
   d. systematic desensitization

Answer: a

3. **According to the AE article, what is the most important thing in preventing failure in relationships?**
   a. money
   b. forgiveness
   c. communication
   d. love

Answer: b

4. **Positron emission tomography (PET) has indicated that schizophrenics have**
   a. diathesis stress.
   b. less activity in the prefrontal lobe.
   c. increased metabolic activity in the frontal lobe.
   d. smaller ventricles in the brain.

Answer: b

5. **Which of the following characters would be most likely to be diagnosed with disthymic disorder?**
   a. Tim "The Tool Man" Taylor
   b. Kramer
   c. Oscar the Grouch
   d. Daffy Duck

Answer: c

6. **Your gruesome, beastly jailer walks in and offers you the choice of being whipped or clubbed as your punishment of the day. Assuming that neither alternative appeals to you, we could assume that you are experiencing**
   a. double conflict.
   b. avoidance-avoidance conflict.
   c. approach-avoidance conflict.
   d. approach-approach conflict.

Answer: b

7. **In a wide-scale study of mental disorders among Americans, about ____ of the sample reported having had at least one mental disorder during their lifetime.**
   a. 25 percent
   b. 50 percent
   c. 10 percent
   d. 35 percent

Answer: b

8. **Biofeedback trains people to**
   a. lower the level of epinephrine in their bloodstream.
   b. exert more control over their autonomic nervous system.
   c. make effective use of defense mechanisms.
   d. use emotion-focused coping.

Answer: b

9. Deinstitutionalization was triggered by
a. the discovery of antipsychotic drugs.
b. reduced funding of inpatient mental hospitals.
c. the increased popularity of electroconvulsive therapy.
d. ethical concerns of abuse in mental institutions.

Answer: a

10. The tendency to overestimate the importance of dispositional factors and underestimate the importance of situational ones is known as
a. schema-driven processing.
b. fundamental attribution error.
c. stereotyping.
d. prejudice.

Answer: b

11. Which of the following would be an example of an event script?
a. thinking that dentists cause people to feel pain
b. waiting in line for movie tickets
c. not renting an apartment to a lesbian couple
d. knowing that all chess players are smart

Answer: b

12. Which of the following ways of reacting to a situation is most likely to increase heart problems?
a. getting angry and yelling
b. taking a deep breath and counting to 10
c. letting it go and not responding to it
d. finding something to distract you

Answer: a

13. According to McCrae and Costa, your personality is most likely to change during what part of the life cycle?
a. after age 30
b. between ages 30 and 50
c. before age 30
d. between childhood and adolescence

Answer: c

14. Which of the following is not one of the five groups of traits in the five-factor model used to describe personality?
a. intelligence
b. agreeableness
c. extroversion
d. neuroticism
Answer: a

15. The physiological changes associated with the fight-or-flight response are caused by the _______ nervous system.
a. somatic
b. sympathetic
c. central
d. parasympathetic
Answer: b

16. Ph.D. is to clinical psychologist as M.D. is to
a. social worker.
b. counseling psychologist.
c. industrial/organizational psychologist.
d. psychiatrist.
Answer: d

17. Why is a relatively new antidepressant, called Prozac, such a popular new treatment for depression?
a. It has fewer side effects than many antidepressants.
b. It prevents episodes of mania.
c. It lowers the level of monoamines.
d. It is more successful in curing depression.
Answer: a

18. When Rachel left for college she promised her parents that she would not use any drugs. At a party someone passed her some pot, which she decided to try. If her superego is operating, this behavior will cause Rachel to feel
a. sadness.
b. pleasure.
c. anger.
d. guilt.
Answer: d

19. Electroconvulsive therapy requires which of the following?
a. physician's decision only
b. written consent, which cannot be withdrawn
c. written consent of patient, or in some cases a legal guardian
d. payment in advance
Answer: c

20. Id is to ego as
a. need is to drive.
b. pleasure is to reality.
c. self is to other.
d. trend is to desire.
Answer: b

21. Shy extroverts, or people who are privately shy, react in which of the following ways to a public situation?
a. They avoid public situations at all costs.
b. They are very quiet, nervous, and withdrawn.
c. They tend to overreact and make fools of themselves.
d. They seem very calm, cool, and collected.
Answer: d

22. Psychoanalysis was developed by
a. John Watson.
b. Sigmund Freud.
c. Carl Rogers.
d. Albert Ellis.
Answer: b

23. In Leon Festinger's boring task experiment, the subjects who were paid only $1 to tell other students it was interesting (a lie) dealt with their cognitive dissonance by
a. insisting that they should also be paid $20 for telling the lie.
b. hoping the students they lied to realized it was just part of the experiment.
c. begging Festinger and his assistants not to reveal their names.
d. convincing themselves that it was somewhat interesting after all.
Answer: d

24. Jennifer has been diagnosed with schizophrenia. She has experienced negative symptoms. Her psychiatrist believes she has little chance of recovery. Jennifer must have

a. unipolar schizophrenia.
b. Type II schizophrenia.
c. Type I schizophrenia.
d. Type III schizophrenia.

Answer: b

25. The part of the brain that is initially activated when a situation is appraised as threatening is the

a. occipital lobe.
b. hypothalamus.
c. cerebellum.
d. adrenal medulla.

Answer: b

26. Freud described developmental stages as periods during which the individual

a. seeks pleasure from different areas of the body.
b. resolves psychosocial conflicts.
c. deals with real-life problems.
d. matures, eventually becoming self-actualized.

Answer: a

27. Psychosomatic symptoms are

a. real physical symptoms that are caused by defense mechanisms.
b. symptomatic of Type B personalities.
c. real physical symptoms that are caused by psychological factors.
d. imagined physical ailments.

Answer: c

28. Mental disorders, according to the DSM-IV, are assessed based on

a. three types of psychological disorders.
b. five separate dimensions.
c. the statistical frequency of abnormal behavior.
d. neurological functioning.

Answer: b

29. The area of behavioral genetics suggests that genetic factors

a. set a range for behaviors.
b. do not interact with the environment.
c. are fixed.
d. account for more than 75 percent of the development of personality traits.

Answer: a

30. Based upon findings of research, which is the best way to cope with an upcoming exam?

a. "There's not much I can do about my grade, so why worry."
b. "I can demonstrate to my professor that I really do know this material."
c. "I wish I could have had more study time."
d. use emotion-focused coping.

Answer: b

Short Answer

Answer two of the following (5 points each).

> **The short-answer and essay questions are generally over subjects that we spent a substantial portion of class time discussing. As I told my class, if one concept or theory filled up the board, it is a good candidate for a major question.**

31. Give an example of a projective test and of an objective test. What are the advantages and disadvantages of each type of test?

Answer:

The Thematic Apperception Test (TAT) is a projective test. It is good because there are no correct answers for the test taker to figure out. It provides unique information about him or her. The disadvantage of this is that the responses may be ambiguous, and the results of the test depend on the interpretation of the therapist.

The MIMPI-2 is an objective test. This test is good because it is relatively quick and easy to administer and score, but the test taker may be able to figure out the "correct" answer, and the test may not be completely accurate as a result.

> **Any examples will do, as long as they are actually examples of the respective tests. At least two specific examples of each type of test are covered both in class and in the textbook, and the advantages and disadvantages are presented and discussed as well. The advantages and disadvantages are fairly specific. It is not necessary to explain each type of test.**

32. What are the three ways we discussed to define abnormal behavior? Briefly give strengths and weaknesses of each method.

Answer:

Statistical frequency is good because it is a very objective method. However, just because something is rare does not necessarily mean it is undesirable. Social norms are good because they are a fairly accurate measure of what is deemed acceptable by the society in which a person lives. However, social norms tend to change over time and may not accurately reflect what is healthy or unhealthy. Maladaptive behavior focuses on the health of the individual and others who may be affected by the behavior. However, this may also be a subjective measure in that there may be disagreement as to whether or not something is good or bad and whether or not that person has a right to act some way, even if it is bad for him or her.

The name of each technique is required. No explanation is necessary. The strengths and weaknesses are fairly specific, but as long as the student gets the point across to me, the answer is fine. The correct answer simply identifies each technique and gives the strength and weakness of each.

33. Briefly explain the Oedipus complex (Electra complex in females). Tell in what psychosexual stage it occurs, what defines the complex, and how it is resolved.

Answer:

The Oedipus complex occurs in the phallic stage and is when a boy develops a fondness for his mother and a dislike for his father. It is resolved when the boy realizes that he cannot defeat his father and so decides to try to be like him so the mother will like him more.

A fairly vague answer is acceptable here, as long as it is clear to me that an underlying understanding of the concept exists. Identification of the phallic stage is required. Identify, even very simply, what the complex is. Also identify how it is resolved. The passage of time is not an acceptable description of the resolution of the conflict.

Essay

Answer the following (10 points).

Answering *the questions asked* is very important on the short-answer and essay questions. When I grade the exams, I assign points for each specific bit of information I ask for. On a short-answer question, for example, some students may fill a page but not answer the question. Another may respond to what was actually asked in two or three sentences. The latter will get more points.

34. When you are in a social or group situation, you are likely to behave differently than you would if you were alone in the same situation. Identify such a situation and name three specific social psychological influences that might affect your behavior. Explain each influence and tell how it operates on your behavior.

Answer:

If I am driving down the road and notice someone having trouble changing a tire at the side of the road, there are several things that may make me less likely to stop. The bystander effect states that the more people there are in a group, the less likely anyone is to help. If there are a lot of cars going by on the road, I am less likely to stop and help. Deindividuation states that in a group situation, people are likely to behave differently in a group than they would as individuals, showing less inhibition and responsibility. Even though I would normally stop and help a person, if everyone else is driving by, I probably won't care enough to stop.

In-group bias is when someone identifies with a group of people, and they are less likely to help people they see as part of another group. I may consider the person changing the tire to be part of an out-group because they are not one of the people driving along with no problem. I might consider the person as part of a group of people who don't take very good care of their cars. As a consequence, I won't want to help them because I don't want to be associated with that group.

A correct response to this question will include any type of group situation, the name of three social psychological influence concepts, and how each of those concepts may influence the student's behavior in that situation. The name of the influence must be given, as well as an explanation of the concept, not just how it will affect the student's behavior. Even general concepts, such as conformity or authority, are acceptable if they are explained in such a way as to illustrate clear understanding and differentiation from the other concepts. Personal reasons or concepts that do not relate directly to social psychology will lose points. How the concepts operate on the student may deviate from the rest of the crowd (e.g., when a person trained in first aid is aware of the social influences and works against them).

Comprehensive Section

Answer all of the following (50 points total):

Try to relate the material to something that interests you. This is essential for the comprehensive part of the final.

1. **Of the psychologists we discussed this semester, who is your favorite and why? (Be specific.)** (5 points)

Answer:

My favorite psychologist is Skinner because he identified very concrete things that we could measure and very practical ways that we could use psychology.

The minimum correct answer here would require giving a psychologist's name and enough information so that I think the student is familiar enough with that psychologist's work to identify it. A more elaborate answer might include more detail on the theorist and perhaps some critical response to his or her work.

2. **Apply three of the psychological theories we discussed to your own academic major or to your life in general. Briefly describe each theory and explain how it would be useful.** (10 points each)

Answer:

I can use the theory of operant conditioning with my son. Operant conditioning identifies four reinforcement techniques, positive and negative reinforcement and positive and negative punishment, that can be used in response to certain behaviors to increase or decrease the behavior. I can use this theory with my son to identify good and bad behaviors that I want to either increase or reduce and use appropriate consequences to modify the behaviors.

Facial feedback theory says that your brain detects the activity of certain facial muscles and interprets those to determine emotions. I can use that to influence my emotions, especially by making myself smile a lot when I feel depressed.

Piaget's theory of development says that we pass through four stages of cognitive development: sensorimotor, preoperational, concrete operational, and formal operational. I can apply this theory to my teaching by understanding that some younger students may not have reached the formal operational stage and may therefore not be able to think abstractly about some concepts very well.

Each answer to this question requires the identification of a theory, at least a brief summary of the theory, and a description of the application of the theory to the student's own situation. The more specific the theory the better. Very broad approaches to psychology are generally not acceptable, unless specific aspects of the theories are applied to specific situations. Think of how you can make practical use of some psychological theories in your own life.

3. If you were to become a psychologist, in which area would you specialize and why?
(5 points)

Answer:

I would want to be a physiological psychologist because I think that physiology is the root of all behavior, and by understanding physiology, I will have the best basis for understanding behavior. I can always build on my knowledge with other perspectives.

Simply identify a research perspective or psychological approach and tell why it interests you in particular. Saying generally that you want to be a social worker or a psychiatrist will not get full credit. Those are not areas of psychology, per se. Think of an area (many are covered in the first couple of weeks of class) that interests you and explain why.

4. If someone were to tell you that psychologists have *proved* that children who are spanked will have psychological abnormalities, how would you respond? (I am interested in your response to the person's attitude about psychology, not development specifically.)
(5 points)

Answer:

I would probably ask how they were able to prove such a relationship. Psychology cannot really prove things absolutely; it can find support for theories or concepts. There seem to be a lot of variables involved in such an issue to make such a strong statement.

Read the whole question and respond accordingly. Responses that give a personal attitude toward spanking are not worth much. If the answer addresses spanking as an example, that is fine. The attitude toward psychology is what is crucial. A correct answer will address the flawed nature of the way of thinking about psychology that is demonstrated in the question. It will show critical thinking and a skepticism toward oversimplifying complex subjects. Explain why this perspective is not a strong one.

5. If someone asks you (kind of like I'm doing now) why psychology is important (or why someone should take Psychology 110), what will you tell them?
(5 points)

Answer:

Psychology impacts almost every aspect of your life in some way. Psychology 110 gives you kind of a sampling of all the things that go on in psychology that are trying to increase understanding of behavior.

This question is fairly open, but answers are remarkably similar to the example given (although usually substantially longer). Tell what you have gotten out of the course and what you have learned from psychology. Rationalize why a university would do something like require you to take a course such as introductory psychology. It is difficult to miss points on this question, but an answer that shows closed-mindedness and failure to see the practicality and general applicability of psychology would do it.

CHAPTER 24

KANSAS STATE UNIVERSITY

PSYCHOLOGY 110: GENERAL PSYCHOLOGY

Guy Vitaglione, Instructor

FIVE EXAMS, ALL MULTIPLE CHOICE, ARE GIVEN IN THIS SECTION OF PSYCHOLOGY 110, each of which accounts for 20 percent of the course grade. Examples of one midcourse exam and the final are given here.

The course is intended to create awareness of the breadth of psychological fields and of psychology as a science, an awareness of the connection between personal experiences and psychological science. I evaluate students on their demonstration of awareness of and appreciation of psychological scientific approaches to human behavior, and recognition that there is a plurality of perspectives on a single given behavior. Accordingly, the exams test the students' knowledge of the text and class material, and their ability to apply their knowledge to specific, concrete examples.

Students should read the course text carefully and thoroughly and ask questions in class as well as participate in class discussions. Not only should students take careful note of what is discussed in class, but they should pay particular attention when what is discussed overlaps with material that is already presented in the text. Students can count on such material reappearing on an exam.

MIDTERM EXAM

Circle the letter corresponding to the best response to each question.

Multiple Choice

1. **More boys than girls have academic difficulty in the first grade. How would a developmental psychologist probably explain this difference?**
   a. Boys are more vulnerable to emotional disorders.
   b. Teachers and parents have different expectations for boys as compared to girls.
   c. Boys mature more slowly than girls.
   d. Girls have two X chromosomes and boys have only one.

Answer: c

The question tests understanding of the perspective adopted by developmentalists. Although several other choices are factually correct, the successful answer reflects an ability to distinguish differing perspectives on psychology.

2. **Who of the following would be doing something illegal or unethical?**
   a. a psychiatrist who makes an appointment with someone who is not mentally ill
   b. a clinical psychologist who prescribes tranquilizers for an emotionally disturbed patient
   c. a psychoanalyst who claims to be a medical doctor
   d. a psychiatrist who signs papers to discharge someone from a mental hospital

Answer: b

Like question 1, this item attempts to discern whether students have understood the basic psychological professions and their individual characteristics.

3. **For some groups of people, the greater the level of stress, the greater the probability of a heart attack. This relationship is an example of**
   a. a positive correlation.
   b. a negative correlation.
   c. a zero correlation.
   d. significance.

Answer: a

All of the answers relate to correlation statistics. The correct answer indicates that the student has not only learned the basics of correlations, but also can accurately apply that knowledge to a concrete situation.

4. **In order to understand the unusual behavior of an adult client, a clinical psychologist has carefully investigated the client's current life situation and his physical, social, and educational history. Which research method has the psychologist employed?**
   a. survey
   b. case study
   c. experimental research
   d. naturalistic observation

Answer: b

All of the answers are research methods, but the question tests whether students have learned the material and if they have sufficiently incorporated their learning to apply it to specific examples.

5. **The hindsight bias refers to people's tendency to**
   a. brag about their past academic success.
   b. exaggerate their ability to have foreseen the outcome of past events.
   c. criticize the research and theories of professional psychologists.
   d. insist that they understand themselves better than any psychologist ever could.

Answer: b

This simply tests factual knowledge. Answers a and d appear to be somewhat melodramatic, and students should be able to detect them as "distractors," false answers, leaving only b and c to choose from.

6. **Teresa and Deborah are subjects in a psychological experiment. The experiment is concerned with increasing self-esteem by listening to classical music. Teresa is asked to sit in a room for an hour listening to classical music. Deborah is asked to sit in a room for an hour listening to no music. Deborah is in the**
   a. experimental condition.
   b. control condition.
   c. observational condition.
   d. sucker condition.

Answer: b

This item again tests the ability to apply factual class information to a simple situation. Note that d is a throwaway answer, reducing the possible correct responses to three.

7. **Subjects in experiments are randomly assigned to different conditions in order to**
   a. make the conditions roughly unequal in terms of subjects' age, race, opinion, or any other preexisting factor.
   b. conceal from the subjects which condition they are in.
   c. make the conditions roughly equal in terms of subjects' age, race, opinion, or any other preexisting factor.
   d. increase the effect of the dependent variable on all conditions equally.

Answer: c

Note that a and c are almost identical. Students should get the hint that the correct answer may be either a or c, since half of the possible answers deal with the same issue.

8. **As scientists, psychologists adopt an attitude of skepticism because they believe that**
   a. people are unlikely to reveal what they are really thinking.
   b. most commonsense ideas about human behavior are wrong.
   c. ideas about human behavior need to be objectively tested.
   d. all of the above

Answer: c

This item tests whether students have incorporated different pieces of information to form a mental picture of how a psychologist thinks and acts.

9. **Michael has been recording how many times each day his father complains about the New York Giants' poor football record. Michael is conducting**
   a. a case study.
   b. an observation.
   c. an experiment.
   d. a survey.

Answer: b

All answers here are relevant to research, and students sometimes mistakenly indicate that a or d is the correct answer: a, because it focuses on one subject, and d, because the researcher is making notes. But a refers to a thorough examination of everything about the subject, and d typically involves talking with the subject. It is important to read the question carefully in order to understand what Michael did and did not do. He *recorded*. We are not told that he made a thorough examination or interviewed the subject.

10. Replication is often viewed as the most important part of the scientific method because

a. an experimental finding is considered significant until proven otherwise.
b. observations must be repeated before they can be examined.
c. the result of an experiment is not accepted as fact until other scientists can repeat the experimental result themselves.
d. experiments with heredity involve multiple conditions.

Answer: c

This is an issue I discuss at length in class. It is not highlighted in the class textbook; therefore, good class note taking is essential to answering this item correctly.

11. Of the major perspectives in psychology, which one examines the effect of brain and gland chemicals on behavior?

a. social
b. clinical
c. psychiatric
d. biological

Answer: d

This one is simple and direct, tapping factual knowledge. It is one of the easier questions, since merely knowing what the words mean can strongly suggest the correct answer.

12. Janet is interested in how temperature can affect our sleepiness. She designs an experiment in which some subjects are instructed to sit in a room that is quite warm, while other subjects sit in a room that is quite cool. She then watches the subjects in each room for signs of sleepiness. In this experiment, the independent variable is

a. time.
b. sleepiness.
c. temperature.
d. observation.

Answer: c

Taps whether students remember the text and class material, whether they understand the material, and whether they can apply what they know. I usually warn students that such "applied questions" will be on the exam.

13. Psychology is best described as the study of

a. the mind.
b. behavior.
c. both a and b
d. the interaction between nature and nurture.

Answer: c

A basic factual knowledge item. Both a and b are obvious, letting the student decide between c or d.

14. A group of friends decided to keep track of how many books they had read during the past year. Here is what they found:

Name	# of books read
Martha	2
Victoria	2
Carolyn	9
Anna	12
Laurel	25

The mean number of books read by these individuals was

a. 2.
b. 9.
c. 10.
d. 50.

Answer: c

Demands math skills and knowledge of what *mean* is. Students were taught mean, median, and mode, and how to distinguish among them. I warned them ahead of time that I would provide simple math problems on tests for this information. Note that the median and modal answers are available as distractors. The students must know the correct definition of *mean* for a correct answer.

15. When reading a histogram, you should always

a. count how many bars it has.
b. look to see if the bars are color-coded.
c. look at its scale.
d. count how many hists are in a gram.

Answer: c

This is part of basic table-reading information I provide in class and in text. The question calls for recall of factual knowledge.

16. When Mr. Ricks calculated his students' algebra test scores, he noticed that two students had extremely high scores. Which measure of central tendency is affected most by the scores of these two students?
   a. mode
   b. median
   c. mean
   d. range

Answer: c

This is a difficult question, because the answer relies on the "central tendency" part of the question. Both c and d are correct answers, but d is not a measure of central tendency; therefore, the answer must be c. The question requires careful reading and then recall of very precise factual knowledge.

17. A correlation between two factors can tell us only that
   a. one factor is caused by another factor.
   b. a relationship exists between two factors.
   c. the two factors are caused by a third, unstudied factor.
   d. it is probably an illusory correlation.

Answer: b

Taps rote-learned factual knowledge. Note that a and c are valid, but they are not precisely correct for this question.

18. Colette received an unusually high grade of A on her first biology test and a B+ on the second test, even though she studied equally for both tests. Which of the following best explains Colette's deteriorating pattern of performance?
   a. the standard deviation
   b. illusory correlation
   c. the random sampling effect
   d. regression toward the mean

Answer: d

Another difficult math item, requiring ability to distinguish between different types of statistical phenomena. They all exist, but d is the only one that matches the question precisely.

19. After she was painfully deceived by her boyfriend, Mary concluded that men just can't be trusted. In this instance, Mary ought to remind herself that reasonable generalizations often depend on

a. an awareness of regression toward the mean.
b. the study of representative samples.
c. maximizing the standard error of the mean.
d. positive correlations.

Answer: b

Answers c and d are patently wrong. Students who eliminate them are left with a or b. If they know anything about these two choices, they can guess the correct answer. This is another item calling for applying definitional knowledge to a concrete situation.

20. A statistically significant difference between two sample groups is *not* likely to be

a. repeatable.
b. a reflection of differences between the populations they represent.
c. observed when the groups are very large.
d. due to random variation within and between the sample groups.

Answer: d

The other three answers are distractors. Memorization of what "statistically significant difference" is should point to the right response. The other answers do exist and are relevant, but are not the direct definition.

21. Sara Lynn, the youngest child of a high school athletic director, was able to roll over at 3 months, crawl at 6 months, and walk at 12 months. This ordered sequence of motor development was largely due to

a. maturation.
b. cultural norms.
c. parental encouragement.
d. imprinting.

Answer: a

All the answers exist and deal with developmental issues, but not with the information provided in the example. That Sara Lynn's father is an athletic director may evoke a commonsense response of c, but this would be jumping to an unwarranted conclusion, since the information in the question does not say anything about "parental encouragement." Only a knowledge of each of the possible answers, especially what constitutes maturation, will yield the correct response.

22. **Tom and Amelia are two developmental psychologists having an argument. Tom claims that people have traits and dispositions that are largely permanent, while Amelia claims that traits and dispositions alter over time. Tom and Amelia are arguing over the**
    a. nature vs. nurture issue.
    b. continuity vs. stage issue.
    c. Kohlberg vs. Gilligan issue.
    d. change vs. stability issue.

Answer: d

This subject is discussed in detail in class. All answers were discussed and are relevant, but only d is pertinent specifically in this case. The question requires both recollection of what each issue is and the correct application of the information to the example provided.

23. **Frances expected all college professors to be old, bearded males. She found it difficult to accept young Susannah Ticell as a legitimate professor due to her own restrictive**
    a. concrete operational intelligence.
    b. egocentrism.
    c. temperament.
    d. schema.

Answer: d

The question deals with cognition. Response c is a throwaway, leaving the remaining three cognitive answers. Inadequate understanding of the three remaining answers would make the question difficult.

24. **Babies born with fetal alcohol syndrome are proof that alcohol is**
    a. a gene.
    b. a form of DNA.
    c. a neurotransmitter.
    d. a teratogen.

Answer: d

This is a simple definitional question. Common knowledge of the nature of alcohol would eliminate answers a and b.

25. During the course of successful prenatal development, a human organism begins as a(n) _______ and ends up as a(n) _______.
    a. embryo; zygote
    b. zygote; fetus
    c. zygote; embryo
    d. fetus; embryo

Answer: c

This question asks for knowledge of a chronology of events. If the full chronology is not known, at least a partial knowledge could help eliminate some responses. For example, knowing that the zygote comes first would eliminate a and d. The question emphasizes knowledge of the chronology over knowledge about the individual stages.

FINAL EXAM

"Distractors" are the incorrect answers that are included in multiple-choice exams to force test takers to be very careful in their selections. An effective approach to a multiple-choice question is to begin by eliminating the distractors—especially the obviously impossible options.

Most multiple-choice questions call for recognition and recall of material presented in a textbook or in class discussion and lecture. If you recall the material accurately, you will respond correctly. However, even if your recall is shaky, you should be able to make at least an educated guess at a correct response:

- **Eliminate the impossible responses.**
- **Scan the choices provided for something that at least *looks* familiar.**
- **Scan the choices for responses that *sound* correct.**

The so-called "stem" of a multiple-choice question—the part containing the information on which the question is based—contains key words valuable for answering the question. Make certain that you understand what the question is looking for by ensuring that you recognize the key words. Circle these, if necessary. Above all, read the questions carefully. Many incorrect responses are due to simple misreading.

Watch out for absolutes. If a stem contains such words as *always, never, none,* or *all,* you must ensure that the response you choose does not mean *often, rarely, some,* or *sometimes.* Writers of multiple-choice exams often create trick questions that are designed to trap the unwary with such absolutes.

Finally, do the easy questions first. Make certain that you answer *all* of the questions you are certain of. Get credit for those. Then go back and tackle the questions that give you trouble.

Circle the letter corresponding to the best response to each question.

Multiple Choice

1. Gestalt psychologists emphasized that
   a. perception is the direct product of sensation.
   b. we learn to perceive the world through experience.
   c. the whole is the sum of its parts.
   d. we organize sensations into meaningful patterns.

Answer: c

Definitional knowledge is required here. Answer a is patently false, leaving three answers, each of which is true, but only c directly applies to Gestalt. The item requires knowledge of all three answers in order to distinguish among them.

2. Because Mark, Jacob, and Angus were all sitting at the same bowling lane, Eric perceived that they were all members of the same bowling team. This best illustrates the principle of
   a. proximity.
   b. closure.
   c. continuity.
   d. binocular cues.

Answer: a

3. Almost all the birds in the yard were brown, and the rest were bright red, so Eve perceived them as two distinct groups of birds. This best illustrates the principle of
   a. proximity.
   b. closure.
   c. similarity.
   d. connectedness.

Answer: c

In questions 2 and 3, all answers exist as perceptual phenomena. The student must be able to apply his or her knowledge to the specific example in order to make the right choice.

4. Psychologists are skeptical about the existence of ESP because
   a. there is not a single scientific experiment that yields results supporting the existence of ESP.
   b. very few people believe in ESP or are willing to admit they have experienced it.
   c. researchers have never been able to replicate an ESP phenomenon.
   d. the close relationship between the mind and the brain suggests that ESP is unlikely.

Answer: c

Both a and d are factually incorrect (note that a includes an absolute—"not a single"—which is always a warning to think twice before choosing it), leaving b and c. As I do with my midterm, I warn students that material covered both in the text and in class is extremely likely to appear on the exam. This question addresses material discussed in class rather than in the textbook.

5. In *The Attack of the 50-foot Woman*, we saw that filmmakers used a technique they called "forced perspective." Forced perspective refers to
   a. changing the perceived size of two objects by increasing their relative distance.
   b. using retinal disparity to increase the perceived size of objects.
   c. changing the perceived distance of two objects by increasing their relative size.
   d. giving Daryl Hannah growth hormones.

Answer: a

Answer d is a throwaway, leaving three answers. Response b, retinal disparity, refers to depth perception, and a and c are very similar. Students who rush or don't read closely may make a mistake. Answering this item demands that the student has seen the film and has taken notes on it.

6. Although Camille sees her psychology professor several times each week, she had difficulty recognizing the professor when she walked past her in the grocery store. This best illustrates the effect of _______ on perception.
   a. similarity
   b. context
   c. proximity
   d. connectedness

Answer: b

Simply knowing the common meanings of each word could point the right way. As with most of my questions, this one requires both definitional knowledge and a facility for applying the knowledge in a nonstatic way to a novel situation.

7. Whether we perceive a moving light in the late evening sky as a UFO or as the light of an airplane depends on our
   a. perceptual set.
   b. retinal disparity.
   c. figure-ground relationship.
   d. ability to detect a forced perspective.

Answer: a

A difficult question. I intersperse difficult questions with easy ones to discriminate among the higher- and lower-performing students. Definitional knowledge is required for each response here, but the situation is difficult to apply the knowledge to.

8. Although textbooks frequently cast a trapezoidal image on the retina, students typically perceive the books as rectangular objects. This illustrates the importance of
   a. overlap.
   b. binocular cues.
   c. linear perspective.
   d. shape constancy.

Answer: d

Definitional knowledge is required here.

9. Of the following, which is not a monocular cue for distance?
   a. overlap
   b. texture gradient
   c. retinal disparity
   d. linear perspective

Answer: c

Again, definitional knowledge is called for. Knowledge of any four possible responses helps to eliminate incorrect answers.

10. When airplane pilots attempt to land at an airport shrouded in a hazy fog, the pilots perceive that the landing runway is farther away than it really is. This is caused by
    a. convergence.
    b. aerial perspective.
    c. relative height.
    d. relative motion.

Answer: b

Definitional knowledge is required.

11. After flying from California to New York, Arthur experienced a restless, sleepless night. His problem was most likely caused by a disruption of his normal
    a. REM sleep.
    b. circadian rhythm.
    c. narcolepsy.
    d. sleep apnea.

Answer: b

Responses c and d are disorders, helping to reduce the possible answers to a and b. While a is close, b refers to a pattern (rhythm) that could be disrupted.

12. REM sleep is to intense dreams as stage 4 sleep is to

a. alpha waves.
b. hallucinations.
c. sleep spindles.
d. sleepwalking.

Answer: d

This requires both definitional knowledge and analogous reasoning. The idea here is that thorough knowledge of the material allows for complex reasoning regarding the material.

13. David, Lew, Glen, and Natalie all ordered beer at a bar in Aggieville. The bartender had only two beers left, and he served them to David and Lew. Without telling them, the bartender served Glen and Natalie nonalcoholic beers. After everyone drank their drinks, Glen and Natalie acted just as tipsy and silly as David and Lew. Glen and Natalie's behavior best illustrates

a. that alcohol is a depressant.
b. the effect of nonalcoholic beer on the sympathetic nervous system.
c. the think-drink effect.
d. that drinking can be dangerous.

Answer: c

Answers a, c, and d all exist, but the behavior described doesn't appear to be either dangerous or depressing, leaving c as the correct answer.

14. Breenan was dismayed to discover that some of his rugby teammates were using drugs to enhance their fast footwork and endurance on the playing field. Which of the following drugs were the players likely using?

a. morphine derivatives
b. marijuana
c. amphetamines
d. barbiturates

Answer: c

A correct response requires command of relevant definitions and applied reasoning.

15. Of the following, which is not an explanation for the purpose of dreaming?

a. wish fulfillment
b. problem solving
c. the prevention of apnea
d. maintaining brain stimulation

Answer: c

All answers relate to sleeping. Only knowledge of each of the answers will allow for a proper selection.

16. Brain activity when we are awake most closely resembles brain activity when we are asleep during

a. stage 2 sleep.
b. stage 3 sleep.
c. stage 4 sleep.
d. REM sleep.

Answer: d

Definitional knowledge is required.

17. People are particularly responsive to hypnosis if they

a. strongly believe they can be hypnotized.
b. are below average in intelligence and education.
c. are easily distracted and have difficulty focusing attention.
d. suffer a physical or psychological dependence on alcohol.

Answer: a

Requires definitional knowledge of each response choice.

18. Vance has just witnessed a bank robbery. While the robbers made their getaway, he got a look at the license plate of their car. However, it was so dark he couldn't see the numbers. When police interviewed him, they put Vance into a hypnotic trance to help him remember what he saw. Once hypnotized, Vance instantly gave the police license plate numbers. In this situation,

a. hypnosis relaxed Vance so he could remember the numbers.
b. hypnosis confused Vance, so he deliberately lied about the numbers to please the hypnotist.
c. under the hypnotic state, Vance unknowingly constructed a false memory.
d. Vance was able to remember the numbers due to the post-hypnotic suggestion.

Answer: c

All answers exist. A correct response requires a careful reading of the question, noting that Vance didn't suppress a memory, but simply could not see because conditions were too dark. Then the student must be able to relate knowledge to the specific example.

19. Ernest has just witnessed a bank robbery in which the robbers shot two security guards as they made their getaway. Ernest got a good look at the robbers; however, he was so upset by the violence that he couldn't remember anything he saw when he was interviewed by the police. When he was put into a hypnotic state, he immediately told the police what he saw. In this situation,

a. hypnosis relaxed Ernest so he could remember what he saw.
b. hypnosis confused Ernest so he deliberately lied about what he saw to please the hypnotist.
c. under the hypnotic state, Ernest unknowingly constructed a false memory.
d. Ernest was able to remember what he saw due to the post-hypnotic suggestion.

Answer: a

All answers exist. As with the previous question, this one calls for careful reading of the conditions posed by the question and an ability to relate knowledge to a specific example.

20. Under hypnosis, Mrs. Jones is encouraged by her therapist to vividly experience and describe the details of an argument she had with her father when she was a child. The therapist was employing a technique called

a. post-hypnotic suggestion.
b. age regression.
c. dissociation.
d. post-hypnotic amnesia.

Answer: b

A correct response requires problem-solving ability and definitional knowledge.

21. Last year, Dr. Stohler cleaned Erika's skin with rubbing alcohol prior to administering each of a series of painful rabies vaccination shots. Which of the following processes accounts for the fact that Erika currently becomes fearful every time she smells rubbing alcohol?

a. observational learning
b. classical conditioning
c. negative reinforcement
d. operant conditioning

Answer: b

All answers exist, making this a difficult applied knowledge item. Only a thorough familiarity with each answer ensures picking the right response.

22. A real estate agent showed Scott several pictures of lakeshore property while they were eating a delicious, mouth-watering meal. Later, when Scott was given a tour of the property, he drooled with delight. For Scott, the lakeshore property was a(n)
    a. unconditioned stimulus.
    b. unconditioned response.
    c. conditioned stimulus.
    d. conditioned response.

Answer: c

Definitional knowledge is called for.

23. In a psychological experiment, infants were continually presented with books, which were followed by a loud, frightening noise. The infants would cry when they heard the noise. After a while, the infants were presented with just the books. At the sight of the books, the infants would start crying. In this example, the crying became the
    a. unconditioned stimulus.
    b. unconditioned response.
    c. conditioned stimulus.
    d. conditioned response.

Answer: d

Definitional knowledge is required for a correct response.

24. When a conditioned stimulus is not followed by an unconditioned stimulus, the subsequent fading of a conditioned response is called
    a. discrimination.
    b. generalization.
    c. extinction.
    d. spontaneous recovery.

Answer: c

Simple definitional knowledge is called for.

25. Because Martha would always pick up her newborn daughter when she began to cry, her daughter is now a real crybaby. In this case, picking up the infant served as a(n) _______ for crying.

a. negative reinforcer
b. conditioned stimulus
c. positive reinforcer
d. unconditioned stimulus

Answer: c

Definitional knowledge and applied reasoning are required.

CHAPTER 25

NORTHEASTERN UNIVERSITY

PSYCHOLOGY 1111: INTRODUCTION TO PSYCHOLOGY

Nayantara Santhi, Graduate Teaching Assistant

TYPICALLY, TWO EXAMS, A MIDTERM AND A FINAL, ARE GIVEN IN THIS COURSE. THE final is included here. Sometimes an instructor chooses to give a set of four exams, without a cumulative final.

The primary objectives in teaching the course are to:

- Introduce students to the varied subject matter and concepts of psychology.
- Ensure that the students learn the concepts and theories in the major areas of psychology, especially physiological psychology, sensation and perception, learning and motivation, consciousness, social psychology, and personality theories.

I would like the students to be able to understand and think through the various concepts they learn in the course. Apart from factual questions, I want them to be able to answer conceptual and integrative questions regarding the material taught in the course. The course examinations consist of true-or-false and multiple-choice questions. While some of the questions in the exam may involve facts, most of them are conceptual in nature and require a good understanding of the material. Accordingly, students should keep up with the material in class. Usually an introductory course involves a lot of material; therefore, students should learn the material as soon as it is taught in class. They should try to understand the concepts rather than merely memorize facts. I would also advise them to use study guides and do the sample exams in the guides. Students should develop an organized schedule for efficient learning. It should be consistent and systematic rather than cramming before the exam. It is important that students read the text in detail. They should take effective notes during the lecture and use them as a guide to selecting material to focus on.

Most of the questions in an exam involve material specifically covered in the lecture. A very small percentage of the questions involve additional material. This is where a careful reading of the text is an effective strategy.

Preparing for a multiple-choice exam involves surveying the material initially, then reading through it in detail while simultaneously highlighting and learning the important points, then finally reviewing the learned material by giving yourself a mock exam.

FINAL EXAM

True or False

1. **The visual cliff is a glass platform that extends over a several-foot drop off and is extensively used in infant perception research.**
   a. True
   b. False

Answer: True

2. **The cephalocaudal trend is the head-to-foot direction of motor development.**
   a. True
   b. False

Answer: True

3. **Separation anxiety occurs only with mothers.**
   a. True
   b. False

Answer: False

4. **In Piaget's theory of development, assimilation involves interpreting new experiences in terms of existing mental structures without changing them.**
   a. True
   b. False

Answer: True

5. **Authority orientation is characterized by a rigid conformity to society's rules, a law-and-order mentality, and avoiding censure for rule breaking.**
   a. True
   b. False

Answer: True

6. **Repression is keeping distressing thoughts and feelings buried in the unconscious.**
   a. True
   b. False

Answer: True

7. **According to Adler, compensation involves efforts to overcome imagined or real inferiorities by developing one's abilities.**
   a. True
   b. False

Answer: True

8. **Bandura is a proponent of the observational learning theory.**
   a. True
   b. False

Answer: True

9. **In Rogers's person-centered theory, incongruence is the degree of disparity between one's self concept and one's actual experience.**
   a. True
   b. False

Answer: True

10. **Locus of control is a generalized expectancy about the degree to which individuals control their outcomes.**
    a. True
    b. False

Answer: True

11. **In an avoidance-avoidance conflict a choice must be made between two unattractive goals.**
    a. True
    b. False

Answer: True

12. **The general adaptation syndrome is a model of the body's stress response. It consists of three stages: alarm, resistance, and exhaustion.**
    a. True
    b. False

Answer: True

13. One of the brain-body pathways through which stress signals are sent to the endocrine system is routed through the cerebral cortex.

a. True

b. False

Answer: False

14. In approach-avoidance conflict a choice must be made between two attractive goals.

a. True

b. False

Answer: False

15. Defense mechanisms are largely conscious reactions that protect a person from unpleasant emotions such as anxiety and guilt.

a. True

b. False

Answer: True

16. The current DSM system uses a dual-axis system of classification.

a. True

b. False

Answer: False

17. Epidemiology is the study of the distribution of mental or physical disorders in the population.

a. True

b. False

Answer: True

18. A phobic disorder is marked by a chronic high level of anxiety that is not tied to any specific threat.

a. True

b. False

Answer: False

This is tricky. The *perceived* threat is very specific.

19. Somatoform disorders are physical ailments that have no authentic organic basis and are caused by psychological factors.

a. True

b. False

Answer: False

20. In dissociative amnesia, people lose their memory for their entire lives along with their sense of personal identity.
    a. True
    b. False

Answer: False

21. Bipolar mood disorders are marked by the experience of both depressed and manic periods.
    a. True
    b. False

Answer: True

22. Hallucinations are false beliefs that are maintained even though they clearly are out of touch with reality.
    a. True
    b. False

Answer: False

It is easy to confuse delusion and hallucination.

23. Disorganized schizophrenia is characterized by a particularly severe deterioration of adaptive behavior.
    a. True
    b. False

Answer: True

24. Social withdrawal, disorders of speech, abnormal motor behavior, and poverty of speech are some of the symptoms exhibited by people diagnosed with schizophrenic disorders.
    a. True
    b. False

Answer: True

25. Tardive dyskinesia can be reversed if detected in its early stages.
    a. True
    b. False

Answer: True

26. Antipsychotic drugs are used to very quickly reduce psychotic symptoms.
a. True
b. False
Answer: False

27. Cognitive therapists usually make clients relive their traumatic childhood experiences.
a. True
b. False
Answer: False

This is a technique associated with psychoanalytic therapists, not cognitive therapists.

28. Psychoanalysts often encourage transference.
a. True
b. False
Answer: False

Can be a tricky question. Transference is an aspect of the psychoanalytic process, but it is ultimately a stage of analysis that must be worked through; otherwise, it is destructive.

29. Aversion therapy involves pairing an aversive stimulus with a stimulus that elicits a desirable response.
a. True
b. False
Answer: False

30. A spontaneous remission is a recovery from a disorder that occurs without formal treatment.
a. True
b. False
Answer: True

Multiple Choice

1. **Homeostasis is a state of**
a. physiological stability.
b. physiological instability.
c. psychological stability.
d. psychological instability.
Answer: a

2. **One's need for social status and for respect and recognition from others reflect the need for ______, which comes into play only after the needs for ______ are satisfied.**
   a. safety; security and esteem
   b. esteem; aesthetic and love
   c. esteem; love and belongingness
   d. esteem; safety and security

Answer: c

3. **The correlation between hormone levels and sexual activity in a human can be explained by the following:**
   a. Hormonal surges cause sexual arousal.
   b. Sexual arousal causes hormonal surges.
   c. neither a nor b
   d. either a or b

Answer: d

4. **Anxiety tends to ______, because ______.**
   a. increase; people want to share and evaluate their concerns
   b. increase; the mere physical presence of others is comforting
   c. decrease; people like solitude when thinking about an anxiety- provoking situation
   d. decrease; other people may be a source of further anxiety

Answer: a

Take your time. Try out each alternative. Read them to yourself.

5. **The motive to avoid failure**
   a. may stimulate achievement.
   b. may inhibit achievement.
   c. both a and b
   d. neither a nor b

Answer: b

Think about the best answer.

6. **A baby refuses to crawl over the edge of the visual cliff. This demonstrates that the baby**
   a. is able to perceive depth.
   b. has good motor coordination.
   c. is able to crawl.
   d. none of the above

Answer: a

7. **Current research shows that _______ is most fully developed at birth.**
   a. hearing
   b. vision
   c. motor control
   d. sense of smell

Answer: a

8. **Which of the following describes the phenomenon of maturation?**
   a. learned behavior
   b. emotional maturity
   c. genetic programming
   d. cognitive development

Answer: c

In a psychological context, *maturation* is a physiological process. Do not confuse it with *maturity*.

9. **The close emotional bond between an infant and its caregiver is called**
   a. dependency.
   b. imprinting.
   c. identification.
   d. attachment.

Answer: d

10. **In developing his theories of psychosocial stages, Erikson revised and built upon the theories of**
    a. Freud.
    b. Piaget.
    c. Kohlberg.
    d. Maslow.

Answer: a

11. **When we describe someone as quiet and unassuming, we are describing their**
    a. unconscious motives.
    b. personality traits.
    c. philosophy of life.
    d. perception of self.

Answer: b

12. Jane has devoted her life to the search for pleasure and immediate need gratification. Freud would say that Jane is dominated by

a. her ego.
b. her superego.
c. her id.
d. all of the above

Answer: c

13. Freud's concept of the unconscious is most like Jung's concept of the

a. preconscious.
b. archetypes.
c. personal unconscious.
d. collective conscious.

Answer: c

Be careful to distinguish between a, b, and c.

14. Operant conditioning is to Skinner as _______ is to Bandura.

a. classical conditioning
b. social learning theory
c. genetics
d. self-actualization

Answer: b

15. Rogers's view of personality structure centers around a single construct, and that is

a. ego.
b. the self-concept.
c. the environment.
d. personality traits.

Answer: a

16. Which of the following describes a person most likely to experience frustration?

a. one who can predict the outcome of an event
b. one who is a perfectionist
c. one who sets goals that are too low
d. all of the above

Answer: b

17. The hallmark of conflict is the feeling of

a. indecision.
b. anger.
c. depression.
d. hopelessness.

Answer: a

Think about the term *hallmark*. It means *defining* characteristic— not associated effect.

18. In which of the following cases might the use of a defense mechanism be adaptive?

a. blaming your coworker for your mistakes
b. justifying cheating on your taxes by claiming that the wealthy pay very little in terms of taxes
c. jogging five miles a day to release work-related tensions
d. refusing to admit one's mistakes

Answer: c

Ask yourself what *adaptive* means. This will prompt you to reject all of the alternatives, save c.

19. Overall, the correlation between stress and illness is

a. moderately positive.
b. strongly positive.
c. strongly negative.
d. zero.

Answer: a

20. Which of the following personality types is most resistant to stress?

a. the hot reactor
b. the pessimist
c. the hardy personality
d. the type A personality

Answer: c

21. Dysthymic disorder is to _______ as cyclothymic disorder is to _______.

a. major depression; unipolar disorder
b. manic depression; unipolar disorder
c. major depression; bipolar mood disorder
d. manic depression; bipolar mood disorder

Answer: c

22. A person who perceives stimuli that aren't there is said to have

a. hallucinations.
b. delusions.
c. obsession.
d. illusions.

Answer: a

Be careful to distinguish a from the other alternatives.

23. Hallucinations are disturbances of

a. communication.
b. emotion.
c. perception.
d. motor behavior.

Answer: c

24. Harry is excessively orderly, neat, rigid, efficient, and inhibited. He constantly frets about having insufficient time to get things done despite careful schedules and plans. He is most likely to receive a diagnosis of _______ personality disorder.

a. paranoid
b. obsessive-compulsive
c. antisocial
d. passive-aggressive

Answer: b

25. Which of the following is not a characteristic of the antisocial personality?

a. manipulative behavior
b. social charm
c. excessive guilt
d. aggressiveness

Answer: c

The antisocial personality is incapable of guilt.

26. According to Rogers, personal distress occurs when

a. unconscious conflicts threaten to rise to the surface of conscious awareness.
b. a person engages in negative thinking.
c. there is an incongruence between a person's self-concept and reality.
d. the person is lacking in self-control.

Answer: c

27. After undergoing psychoanalysis for several months, Jane has suddenly started forgetting to attend her therapy sessions. Jane's behavior is most likely in the form of

a. resistance.
b. transference.
c. insight.
d. catharsis.

Answer: a

Think about Jane's behavior. Clearly, she is *resisting*.

28. The key task of the client-centered therapy is

a. interpretation of the client's thoughts, feelings, memories, and behaviors.
b. clarification of the client's feelings.
c. confrontation of the client's irrational thoughts.
d. modification of the client's problematic behaviors.

Answer: b

29. Cognitive therapy emphasizes

a. recognizing and changing negative thought patterns.
b. reliving traumatic childhood experiences.
c. increasing the client's self-awareness and self-acceptance.
d. modifying maladaptive behaviors.

Answer: a

30. Antianxiety drugs

a. can permanently cure anxiety disorders.
b. are prescribed only for people who have a clinical anxiety disorder.
c. can temporarily alleviate feelings of anxiety.
d. all of the above

Answer: c

CHAPTER 26

NORTHERN MICHIGAN UNIVERSITY

PY100: PSYCHOLOGY AS A NATURAL SCIENCE

Alan J. Beauchamp, Associate Professor

THE COURSE INCLUDES A TOTAL OF SIX EXAMS. THE MIDTERM AND FINAL, PRESENTED here, are overnight take-home essay exams.

My primary objectives in teaching this course are to introduce students to the core areas within the field, to show them how we do research, and to get them to think about how the various topics discussed interrelate. I want my students to master basic terminology and concepts, to be able to extend and apply this learning to novel situations, and to show a good understanding of how the different focuses within psychology can be related to produce a better understanding of human behavior and cognition.

To answer the questions in these exams requires mastery of materials and terminology, an understanding of interrelationships between domains of study, and an ability to apply this understanding within the context of the question. Prepare yourself by going to class, reading the textbook, asking yourself questions as you read, and asking in-class questions of the instructor. Try to challenge the instructors' thinking whenever possible. Also, think about how the different areas of psychology relate to one another.

Master materials not only by reading and encoding key terms and passages, but by thinking about how the material relates to your own life. For example: Am I classically conditioned to behave in a particular way in a given situation? Do I selectively forget bad things and remember good things? Also, think about how the domains studied interrelate; for example, how does the study of selective attention relate to the theory of defense mechanisms put forth by Freud?

As for these exams, before you answer the question write an outline to help organize your thoughts.

FINAL EXAM

Provide thorough but concise answers to three of the four questions below.

1. **Contemporary psychologists no longer think about behavior as nothing but a product of either genetic or environmental influences, but instead have come to understand that the specific genotype of the organism will require specific input from the environment for optimal physiological development and self-regulatory capacities to emerge. Discuss how many theories in psychology can be seen as attempts to specify what these genetic needs are.**

Answer:

Most theories in psychology attempt to specify what it is that a human requires for optimal development. Freud discusses how, through a series of maturationally determined stages of development, the environment plays a crucial role in providing appropriate gratification of idinal desires (oral, anal, phallic, genital), the id being a genetically inherited structure.

Erikson discusses the notion that children proceed through a series of maturationally determined stages of development, with each stage representing a particular crisis (e.g., trust versus mistrust). Crisis resolution is a function of environmental input.

Trait theory discusses how some traits are inherited (e.g., physique, temperament, and intelligence), but that the realization of optimal development requires an optimal environment (the example of intelligence).

Humanists like Abraham Maslow very specifically lay out what they believe human needs to be and postulate these needs to be genetically inherited.

Physiological psychology discusses how environment affects neurophysiological development, as illustrated by the work of Hubel and Weisel (and our instructor, of course).

To be successful, a response must: (1) demonstrate a knowledge of the models used to illustrate, (2) demonstrate a knowledge of how genes require environmental support to be expressed properly, and (3) provide an analysis of selected models in terms of required environmental support. This response is successful because it thoroughly discusses how psychology currently views the interaction between nature and nurture. It illustrates an understanding of the models presented and extends this understanding to include a discussion of how each model can be viewed as an attempt to specify environmental needs as they relate to the human genotype.

2. **I want to see if taking vitamins regularly causes a decrease in the number of times students miss lecture. Tell me how you would test this idea.**

Answer:

I would first design the standardized procedure to be used for the study. The independent variable in the study is "regularity" of taking vitamins, the dependent variable is "number of lecture days missed." Both variables need to be operationally defined. Regularity could be operationally defined as taking one multivitamin at noon every day during the observational session, which I could define as weeks 2 through 15 of the semester. A missed lecture could be defined as absence during 30 or more minutes of a 50-minute lecture, and could be quantified by simply counting up the number of absences each student had.

I would next employ control procedures to control for human bias. Subject bias would be the possibility that those receiving vitamins would be more motivated to go to class not because of increased vigor, but to "help" the experimenter get the data he or she wants (or the opposite scenario—skip a lot of classes to foil the experimenter). To circumvent this problem, we would give the experimental group the vitamins and give the control group a placebo, thereby preventing them from knowing whether they have been assigned to the experimental or control group.

Experimenter bias must also be controlled. That is, there is a possibility that research assistants (RAs) may also want to help or hurt the experimenter's chances of obtaining a positive result through biasing the observations. For example, a subject in the experimental group that missed 31 minutes might be scored as "present," but the same subject, if in the control group, might be scored "absent." To control this, two independent observers, ignorant of which subjects are in the control and which in the experimental groups, must be used. In this way we can also determine "inter-rater reliability" to determine if observations are accurate.

The last thing I would need to do before I begin is determine who will be in the experimental and who in the control group. To do this I would randomly assign the students to either the experimental or the control group. At this point, responsibilities of subjects to experimenter (instructions on when and how often subjects are to take their pill) and experimenter to subjects (any instructions and explanation of their ethical rights, of informed consent to participate) would be given. Observations would be made on the dependent variable throughout the targeted time period, and data on missed lectures collected. To determine, in an unbiased way, whether the vitamins reduced the number of missed lectures more than would be expected by chance alone, data would be submitted to statistical analysis.

A successful answer here requires a complete description of the process used to conduct a controlled experiment. Terminology must be presented and adequately used within the context of this discussion. Key concepts that should be discussed include independent and dependent variables, subject sampling and assignment, control over experimenter and subject bias, ethical treatment of subjects, and assessment of collected data. This is a good response, as it thoroughly explains how this experiment would be conducted, includes all pertinent terminology, and demonstrates a good grasp of the experimental approach.

3. **We discussed the notion that there are two classes of research design in psychology, those that allow inference about cause and effect and those that do not. Discuss the strengths and weaknesses of both approaches.**

Answer:

Both causal and noncausal designs are capable of producing excellent information. I do not think of one as being better than the other.

The controlled experiment is the causal design, and the strength of this approach is based upon control. Variables are operationally defined, procedures applied to subjects are standardized, subjects are randomly selected and assigned to experimental and control groups, and methods to ensure accurate, nonbiased observations are employed. In short, as much care as possible is taken to be sure that both experimental and control subjects are treated identically, with the exception of the independent variable. If differences in the means between the two groups are more than would be expected by chance alone ($p \leq .05$), they are assumed to be caused by exposure to different levels of the independent variable. I say "assumed" because even when using the controlled experiment, it is possible that results obtained could have been caused by an unknown third variable—it is, however, much less likely using the controlled experiment. The problem with this approach is that results do not generalize well (i.e., poor external validity). Results cannot be applied with confidence beyond the population from which the sample was drawn or outside of the exact context of the experiment.

Noncausal designs, such as survey, correlation, or naturalistic observation, do not have as much control over the experimental context, the observational process, or the subjects' behavior. More important, they do not incorporate a true independent variable (i.e., although data is often analyzed by age, gender, and so forth, these represent quasi-independent variables and provide no more information than does a correlation). Due to these limitations (i.e., poor internal validity), causal inferences cannot be made. Information gathered does reveal important information on relationships, attitudes, and typical behavior observed within a "real-world" context. Because of this high level of external validity, results are much more likely to generalize, and hence often provide useful applicable information. Further, although these techniques do not permit an understanding of cause and effect, they do provide valuable information that can lead to the development of a hypothesis that can be checked out using controlled experiments.

A good answer requires the student to demonstrate a thorough understanding of causal and non-causal experimental designs in psychology, an ability to discuss the strengths and weaknesses of both approaches, and the understanding of the crucial difference between them (i.e., internal versus external validity). This is a good answer because it is well organized, provides a good description of both designs, and illustrates the crucial differences between them.

4. **Provide a discussion of the interrelationship between the domains of physiological psychology, cognitive psychology, and learning psychology.**

Answer:

All three areas work toward the same goal—an understanding of how human beings can learn so much from their experiences. Learning psychology focuses on the role of the environment in determining how learning occurs and cognitive psychology the role of mental processes in learning, while physiological psychology attempts to understand the role of the CNS in learning.

In the discipline of learning specifically, operant (S-R consequence) and classical (CS-US) conditioning are the primary focus of study. They [behaviorists] have obtained a functional understanding of how environmental stimuli can come to gain control over behavior, but have not (by choice) studied how these stimuli are mentally processed or how they affect mental processes.

The area of cognitive psychology does study the role of mental or "cognitive" processes in learning. Working off meticulously refined methodology provided by the learning psychologists, they began to investigate how humans mentally process information presented. For example, cognitive psychologists discovered how information flows through the human mind (i.e., the processing system), the role of cognitive operations in the encoding, storage, and retrieval of information, and how attention, motivation, and historical experience affected what and how information was processed. In the meantime, physiological psychologists were developing an understanding of both how the brain worked as a whole and how specific parts of the brain relate to behavioral and cognitive control. For example, reading aloud a written word requires involvement of the entire brain, but speech output can be disrupted with damage to Broca's area only. An integration of findings from learning, cognitive, and physiological psychology has been successfully applied in areas such as neuropsychology, clinical psychology, and industrial-organizational psychology.

A good answer here requires a thorough understanding of the three specified areas of study as well as an understanding of how these areas interrelate. This response is a good one. It is well organized and contains appropriate terminology, provides adequate illustration of what the three different models are, and does a good job of discussing how these areas are related to one another.

FINAL EXAM

Each answer must demonstrate mastery of the concepts and terminology required, must demonstrate an ability to extend this understanding to the context of the question, and when required, must reasonably relate to the domains of study targeted in the question.

1. **Sociobiology (aka: evolutionary psychology) has been criticized for taking too deterministic a position regarding the influence of genetics on personality. Is this theory the only one to postulate genetic influence on personality? Is there evidence of the influence of genetics on personality? Provide a discussion of this controversy.**

Answer:

Psychobiology and its proponents (e.g., E. O. Wilson) have been attacked on the basis that their position implies that genes determine personality. This is not, however, what the sociobiological position states. They, as do most theories of personality, simply say that genes have an effect on personality, they do not say genes determine personality—culture and experience interact with genes to determine personality. There is, however, much evidence regarding the influence of genes on personality. For example, evidence for genetic influence can be found in the human universals of territoriality, mating strategies and preferences, nepotism, and the desire to have children.

It is interesting to note that all personality theories, except perhaps radical behavioral theory, postulate genetic influence on personality. For example, Freud provides the "id" as an inherited structure, Jung the "archetype," Allport the "striving for perfection," Cattel the "constitutional source trait," and Rogers the "actualization tendency." Perhaps the strong objections about the notion that genes affect personality are due to the fiasco that occurred when similar claims were made regarding intelligence. In light of the infamous history of the theory that intelligence is determined genetically, such reactions are to be expected, but not to be taken lightly, that is, they certainly do send the message that the lines of communication between science and society need work.

To answer this requires a good understanding of sociobiology, an ability to appraise objectively the data presented by sociobiology, present that data, and discuss how this logic applies to the other models of personality. To do this, of course, requires a thorough understanding of these other models. Furthermore, insight into why the reaction to sociobiology was so negative is helpful.

This is a good response. It is well organized, demonstrates an understanding of the sociobiological position regarding the role of genetics in personality, uses jargon appropriately, and demonstrates a thorough analysis of the other theories of personality to support the response.

2. **In the study of abnormal behavior we found that some speculated that the environment caused the disorder (e.g., unpredictable reinforcement), others the derailment of cognitive capability and processing, others genetic factors, and yet others speculated that the disorder was the result of physiological or biochemical problems. Are these positions at odds?**

Answer:

The short answer is no. Take the example of schizophrenia. First, evidence from behavior genetics suggests a possible genetic influence, as there is a higher concordance rate for schizophrenia between identical versus dizygotic twins. Although this methodology is far from perfect, let us assume that some are more genetically vulnerable to schizophrenia than others. Crucial in determining whether the disorder results (i.e., these genes are expressed) will be the environment. If it meets the need of the individual (e.g., provides predictable and adequate reinforcement), stress will be low, and cognitive and neurodevelopment will proceed in an optimal fashion. As such, the child is much more likely to grow into an adaptive, mentally healthy adult. If the same child is exposed to an environment that does not meet his or her needs, it will be a highly stressful environment, and this will derail optimal neurophysiological development. This will, in turn, lead to problems with cognition and behavioral regulation.

Less than optimal neurological development may also set the stage for the expression of the schizophrenic gene or genes. In this case, the child is more likely to grow into an adult schizophrenic, with the physiological, cognitive, and behavioral indicators of schizophrenia.

To answer this item successfully requires an understanding of how genetics and environment interact to affect physiological development, and hence behavioral and cognitive regulation. To do this requires a synthesis of information from the areas specified.

This response is a good one because it demonstrates a good understanding of these areas and logically relates how these various bodies of information can all be used to explain abnormal behavior—that is, they all have something to contribute and they all work together in a dependent fashion.

3. Relate what we learned in the study of "health psychology" to what we learned in the study of "personality" and "abnormal psychology."

Answer:

Health psychology is primarily the study of how people's life style affects their physical and mental well-being. Health psychologists have outlined various factors associated with risk of mental illness (e.g., the adequacy of social support networks) and physical illness (e.g., alcohol consumption, smoking, inadequate or irregular sleep patterns). Factors associated with affecting behavioral change and coping with illness have also been illuminated. Similarly, the study of personality can be viewed as an attempt to understand human needs, that is, what it is from the environment that the human genome requires for optimal development.

When needs are met, smooth personality development results—if not, maladaptive behaviors are more likely. Many of the maladaptive behaviors discussed by health psychology are clearly related to the negative life-style choices (e.g., abusing intoxicants, smoking, sleep patterns). This, of course, implies that effecting life-style changes will require effecting personality change. Finally, abnormal psychology is the study of maladaptive behavior, its etiology, and its treatment. Through the study of etiology and treatment much has been learned about how maladaptive behaviors are adopted and how they can be treated and replaced with more adaptive substitutes. For example, an understanding of how anxiety disorders are developed and their relationship to maladaptive behavior, such as intoxication and smoking, draws heavily on personality theory, and is directly applicable in the field of health psychology.

To answer this item successfully calls for an understanding of how health psychology is really an application of knowledge gleaned from the study of personality and abnormal psychology. To do this requires an understanding and synthesis of information from these three areas.

This response is a good one because it demonstrates understanding of these areas, and relates how the data of abnormal and personality psychology can be applied within the area of health psychology.

4. Discuss the characteristics of a good psychological test (e.g., a personality or intelligence test).

Answer:

A good psychological test should represent an accurate behavioral sample, whether the focus is on overt or covert behavior. The attribute being measured (e.g., intelligence) should be clearly defined, and the items on the test should adequately sample the domain of the selected attribute (content validity). Items should be clearly written (good short-term test-retest reliability) and be designed so as to show high inter-item correlation (i.e., internal reliability). Items must also be designed to minimize, control, or permit assessment of response biases such as social desirability, yea/nay saying, and lying. If the attribute is postulated to be stable across time, evidence of good long-term test-retest reliability and alternate form reliability should be presented. The test must be designed so that administration is standardized and the scoring of responses is objective and hence reliable. Most important, the test must be valid, or measure what it is supposed to measure. Validity can be demonstrated by showing a correlation between test results and a different measure of the same attribute (convergent validation) or by demonstrating a correlation between test scores and outcome predictions (predictive validation). Other methods to establish validity include internal validation (demonstration of a high inter-item correlation), factor validation (demonstrating that items "load" or correlate together in clusters associated with predefined factors within the test), and content validation (demonstrating an adequate sampling of the domain of the attribute being measured).

Last, to aid in the interpretation of individual test scores, the test should provide a presentation of a representative normative sample.

To answer this question well requires a good understanding of the psychometric features of a good test. Appropriate knowledge and use of jargon will be beneficial. This response is a good one because it shows thorough understanding of all the essential components. For example, reference to standardization, the importance of objective scoring, the control of biased responding, and the properties of reliability and validity are crucial for a successful response here.

CHAPTER 27

WAKE FOREST UNIVERSITY

PSY151: INTRODUCTORY PSYCHOLOGY

Terry D. Blumenthal, Associate Professor

THERE ARE FOUR EXAMS IN THIS COURSE: THREE MIDTERMS AND A FINAL. ONE midterm and a final are included here.

Primary course objectives are to provide a basic knowledge of many areas of psychology, to encourage critical thinking, and to demonstrate the interrelatedness of all areas of psychology. Students need to develop critical reading and thinking skills—the ability to distinguish opinion from fact. They need to appreciate that all information in psychology is related and is cumulative. Accordingly, in the exams, students are required to integrate information from various sections of the course. Later exams assume knowledge from earlier sections of the course.

It is helpful to join a study group. Explain the material to anyone who will listen, and encourage the listener to ask questions and to try to stump you. Read the assigned material before class, then read it again after class.

MIDTERM EXAM

Part I.

Multiple Choice (1 point each). Circle the letter corresponding to the best answer.

1. **In an effort to increase polite behavior in my son, I give him a second dessert for saying "please." From his perspective, this is an example of**
   a. positive reinforcement.
   b. negative reinforcement.
   c. positive punishment.
   d. negative punishment.

Answer: a

2. **Dreams are based on the brain trying to make a coherent story out of random neural activation, according to the**
   a. activation-synthesis theory.
   b. restoration theory.
   c. somnambulism theory.
   d. psychodynamic theory.

Answer: a

3. **While Pavlov was studying learning in dogs in Russia, _______ was studying learning in cats in the United States.**
   a. Thorndike
   b. Skinner
   c. Klappstein
   d. Piaget

Answer: a

4. **Seasonal affective disorder is due to a change in the amount of**
   a. acetylcholine.
   b. dopamine.
   c. endorphins.
   d. serotonin.

Answer: d

5. **The information processing model of memory is based on**
   a. the computer metaphor.
   b. the divided house metaphor.
   c. the hierarchical model.
   d. the imagery model.

Answer: a

6. **A tendency to blame the victim of a crime is likely to occur when someone is trying to preserve his or her**
   a. self-serving bias.
   b. actor-observer effect.
   c. self-handicapping strategy.
   d. just-world hypothesis.

Answer: d

7. **For attributions, situational is to dispositional as**
   a. external is to internal.
   b. internal is to external.
   c. top-down is to bottom-up.
   d. bottom-up is to top-down.

Answer: a

8. **In an effort to get my son to stop pulling our dog's ears, I take away his favorite toy every time he pulls the dog's ears. From his perspective, this is an example of**
   a. positive reinforcement.
   b. negative reinforcement.
   c. positive punishment.
   d. negative punishment.

Answer: d

9. **How did Alfred Binet feel about Walter Stern's summarizing the results of intelligence tests into a single number called the Intelligence Quotient?**
   a. Binet felt that Stern had made an important contribution to intelligence testing.
   b. Binet felt that Stern had done a disservice, since intelligence is too complex to express in a single number.
   c. Binet felt that Stern had introduced cultural bias into his summary score.
   d. Binet felt that Stern focused too much on verbal scores and not enough on performance scores.

Answer: b

10. **Which of the following holds the fewest pieces of information at one time?**
   a. sensory memory
   b. short-term memory
   c. long-term memory
   d. They all hold the same number of items.

Answer: b

11. If you eat all your vegetables, you get a piece of cake. This is an example of the
   a. Discrimination principle.
   b. Premack principle.
   c. Peter principle.
   d. Contingency principle.

Answer: b

12. We are more likely to assume internal causes for the behavior of others and external causes for our own behavior, according to the
   a. stereotypical attribution theory.
   b. fundamental attribution error.
   c. mean-world hypothesis.
   d. dissonance deletion strategy.

Answer: b

13. You find that your studying becomes more efficient if you promise yourself a five-minute break every time you read 10 pages of the text. You have put yourself on a _______ reinforcement schedule.
   a. fixed-ratio
   b. fixed-interval
   c. variable-ratio
   d. continuous

Answer: a

14. A person who suddenly falls asleep in the middle of the day, going straight into REM sleep in less than five minutes, probably has
   a. enuresis.
   b. somnambulism.
   c. narcolepsy.
   d. maintenance insomnia.

Answer: c

15. Pavlov said that language learning is a good example of
   a. desensitization.
   b. second-order conditioning.
   c. operant conditioning.
   d. unconditioned responding.

Answer: b

16. The extent to which people end up doing things that a test would predict they should be good at is called
a. reliability.
b. validity.
c. correlation.
d. inference.
Answer: b

17. According to the theory of ________, your memory will be better if the recall situation is more like the learning situation.
a. self-fulfilling strategies
b. dishabituation
c. encoding specificity
d. comprehension
Answer: c

18. If two ideas are in conflict, they generate tension. Reducing this tension is reinforcing. This is an example of
a. positive reinforcement.
b. negative reinforcement.
c. positive punishment.
d. negative punishment.
Answer: b

19. Regular fluctuations in bodily processes that occur on about a daily cycle are called
a. circadian rhythms.
b. ultradian rhythms.
c. basic rest-activity cycles.
d. diurnal rhythms.
Answer: a

20. We attribute our success to internal causes and our failure to external causes, according to the
a. stereotypical attribution theory.
b. actor-observer effect.
c. just-world hypothesis.
d. self-serving bias.
Answer: d

21. Where does n-acetyltransferase change serotonin into melatonin?
    a. pituitary gland
    b .pineal gland
    c. thyroid gland
    d. adrenal gland

Answer: b

22. Behaviorism grew most directly out of the work of
    a. Freud.
    b. Watson.
    c. Wundt.
    d. Schmidt.

Answer: b

23. A multiple-choice question is an example of which type of memory test?
    a. recall
    b. recognition
    c. relearning
    d. cued recall

Answer: b

Part II.

Short Answer (1 point each). Write your answer in the space provided below the question.

1. We can increase the amount of information held in short-term memory by increasing the amount of information in each item, through a process called _______.

Answer: chunking

2. If an 8-year-old has a mental age of 12, her IQ is _______. Answer: 150

3. _______ is the operant procedure of reinforcing behaviors that are closer and closer to the target behavior. Answer: Shaping

4. Skinner developed a device called the teaching machine in the 1950s. What common device in today's world is most like that teaching machine?

Answer: personal computer

5. The inability to keep breathing while asleep is called _______.
Answer: sleep apnea

6. Leon Festinger developed a theory to describe how people change an attitude to decrease the tension produced by that attitude. This theory is called _______.
Answer: cognitive dissonance theory

7. When we talk about reliability, what statistic are we calculating and reporting?
Answer: correlation coefficient

Part III.

Not-So-Short Answer (2 points each). Write your answer in the space provided below the question.

1. Define the term *mnemonic device*, and give an example.

Answer:

A mnemonic device is anything that can be used to increase memory performance. Examples include visualization, reminder notes, rhymes ("one is a bun," etc.), and organization strategies.

2. What was the specific purpose for which the first intelligence test was developed by Binet and Simon?

Answer:

Binet and Simon were asked by the Paris school system to develop a test that could be used to distinguish between normal and handicapped children. The goal was to separate these two groups, and provide special education for the handicapped children.

3. How does the EEG pattern of REM sleep differ from that of slow-wave sleep?

Answer:

In REM sleep, EEG looks like that during the awake state, with low amplitude, high frequency EEG. In slow-wave sleep, EEG is higher in amplitude and lower in frequency.

4. Define observational learning, and give a detailed example.

Answer:

Observational learning involves a change in the behavior of one person upon watching another person receive reward or punishment for a particular behavior. For example, if I see someone press a button on a wall and receive a soft drink, the probability that I will press the same button may increase, even though I have not yet been reinforced for that button press.

Part IV.

Longer Answer (4 points each). Write your answer in the space provided below the question.

1. Define *contingency of reinforcement,* as described in Skinner's theory. Give me an example of each component of this concept.

Answer:

A contingency of reinforcement has three components: the discriminative stimulus, or situation; the response; and the reinforcer. For example, a teacher asks the class a question, a student raises her hand, the teacher calls on her, she answers the question, and the teacher praises her. From the standpoint of the student, the discriminative stimulus is the teacher asking a question in class; the response is raising the hand; the reinforcer is being called on by the teacher. This then acts as a second discriminative stimulus, with the response being the answering of the question, and the reinforcer for that response being the praise of the teacher.

Students must know the correct terms and be able to define them. They should also know alternative terms for each component, wherever appropriate (e.g., "the discriminative stimulus" is also the "situation"). This gives them experience in finding more familiar names for some of the terms that they learn in this course. They should also be able to supply an example from their own experience, although about half the students describe one of the many examples that I mentioned in class.

A successful answer must correctly list and describe the three components, and must provide at least one example of a contingency of reinforcement.

The best way to solve this problem is in reverse: to think of an example first and then to map that example into the contingency of reinforcement. Of course, this requires that the student have some idea of what the three components are to begin with. The more personally relevant the example, the better. Students begin to see that operant conditioning may explain much more of their learning than they may have thought before this section of the course. Thinking in examples, especially personally relevant examples, helps the student learn the material faster and better.

2. Draw a graph that shows how the conditional response changes as we move through the stages of classical conditioning. Tell me what each of the four stages is called and what stimuli are present, based on the following example: A bell rings just before a door closes on your hand.

Answer:

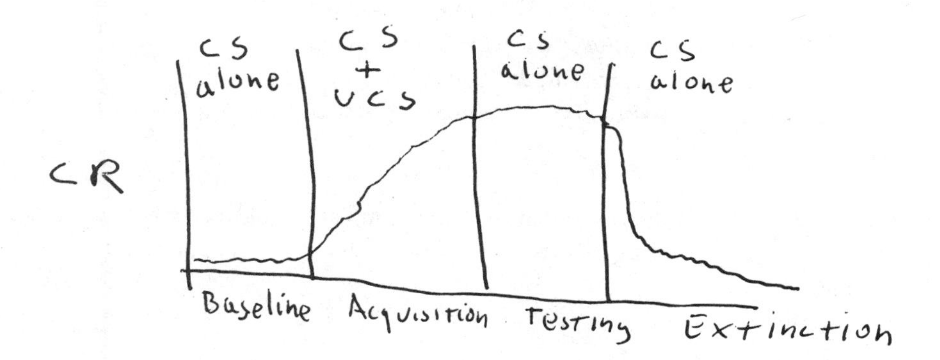

For this answer, the student should draw a graph with conditional response as a function of time. The abscissa will be divided into four sections, labeled baseline, acquisition, testing, and extinction. During baseline, the CS is neutral and is presented alone. During acquisition, the CS and the UCS are paired. During testing and extinction, the CS is again presented alone. We should see a low and stable response to the CS during baseline. This response should increase during acquisition, plateau during testing, and decrease during extinction. Note that the line is not smooth, to demonstrate variability. A smoother line would be acceptable, too.

The student needs to understand the relationships between the CS and UCS, and between the CS and CR, at various stages in classical conditioning. Some students will draw the graph correctly, but not have the names of the four stages correct. Other students will have the names of the four stages, but not have the correct stimulus conditions. Some students confuse CS and CR. A correct answer to this question requires an understanding of the complexity of classical conditioning and the interrelatedness of stimuli and responses, and how these relationships change over time and with exposure.

The answer must include the correct terms for the stages of conditioning, the correct stimulus pairings, and the correct function of CR related to time. The student must think about how a warning can lead to effective behavior, even if the response to that warning is not conscious. The point here is that behavior is changing as stimulus pairing changes, and we have names for the various stages in this process of learning.

A picture is worth a thousand words, and this question evaluates the student's ability to present complicated information in the form of a graph. The student needs to think about how the situation changes across time, what happens to responding when stimulus conditions are changed, and how this change in response can be functionally relevant.

Part I.

Multiple Choice (1 point each). Circle the letter corresponding to the best answer.

1. **Consider these two sentences: "Batman bought a pizza for Robin" and "A pizza for Robin was bought by Batman." These two sentences have**
   a. the same surface structure and the same deep structure.
   b. the same surface structure and different deep structure.
   c. different surface structure and the same deep structure.
   d. different surface structure and different deep structure.

Answer: c

2. **What is the name of the guide that the American Psychological Association has published to help practitioners recognize and diagnose specific disorders?**
   a. DSM-IV
   b. EDR-IV
   c. WHG-IV
   d. DSL-SEM

Answer: a

3. **Seeing other people with problems similar to your own, and learning from the solutions that they use to succeed, is an advantage of**
   a. group therapy.
   b. family therapy.
   c. brief therapy.
   d. cognitive therapy.

Answer: a

4. **Gathering data from people at a range of ages at the same time is an example of**
   a. cross-sequential research.
   b. cross-sectional research.
   c. longitudinal research.
   d. time-lag research.

Answer: b

5. **When do environmental influences begin to have an effect on human development?**
   a. at birth
   b. at conception
   c. in the second trimester
   d. at six weeks of age

Answer: b

6. All of the following are types of communication used in the prespeech stage, except

a. babbling.
b. crying.
c. gestures.
d. telegraphic sentences.

Answer: d

7. A substance that a mother ingests that can harm the fetus is called

a. a neurotoxin.
b. a teratogen.
c. a vomitogenic chemical.
d. an aversive hormone.

Answer: b

8. In Freud's theory, the id operates on the

a. morality principle.
b. pleasure principle.
c. reality principle.
d. defensive principle.

Answer: b

9. Freud felt that sex was the primary motivator of behavior, based on his reading of the work of

a. Wundt.
b. Pavlov.
c. Treichel.
d. Darwin.

Answer: d

10. Panic attacks are often characterized by a sudden increase in the activity of the

a. somatic nervous system.
b. sympathetic nervous system.
c. parasympathetic nervous system.
d. subconscious defense mechanisms.

Answer: b

11. Phobias are an example of
a. a mood disorder.
b. a psychotic disorder.
c. an anxiety disorder.
d. a bipolar disorder.
Answer: c

12. The distinction between introversion and extroversion was proposed by
a. Adler.
b. Freud.
c. Nash.
d. Jung.
Answer: d

13. Which of the following is a treatment for a patient with clinical depression?
a. Thorazine
b. Prozac
c. Lithium
d. L-Dopa
Answer: b

14. Systematic desensitization is best thought of as a form of
a. primal therapy.
b. behavioral therapy.
c. regression therapy.
d. rational-emotive therapy.
Answer: b

15. The process by which unwanted ideas are moved into the unconscious is called
a. regression.
b. repression.
c. depression.
d. digression.
Answer: b

16. Of the following, which is still commonly used in medical treatment of some psychiatric disorders?
a. insulin shock therapy
b. electroconvulsive shock therapy
c. malarial therapy
d. lobotomies
Answer: b

17. The Thematic Apperception Test and the Rorschach Inkblot Test are examples of
    a. projective personality tests.
    b. standardized personality tests.
    c. multiphasic inventory tests.
    d. superego tests.

Answer: a

Part II.

Short Answer (1 point each). Write your answer in the space provided below the question.

1. In Freud's psychodynamic theory, which of the three parts of the personality is what many people now call "the self"?

Answer: ego

2. Kohlberg proposed a stage theory of _______ development.

Answer: moral

3. Which neurotransmitter is unbalanced in many schizophrenics?

Answer: dopamine

4. Finding a less threatening target on which to take out my aggression, such as vandalizing cars when I am angry at my parents, involves which Freudian defense mechanism?

Answer: displacement

5. Who proposed the concept of a personal and a collective unconscious?

Answer: Carl Jung

6. Define *phoneme.*

Answer: the smallest unit of speech

7. Of the three parts of the personality in Freud's theory, which is the only one that has easy access to consciousness and reality?

Answer: ego

8. What development caused the medical community to stop using lobotomies in the 1950s?

Answer: the development of effective psychiatric drug therapy

Part III. Not-So-Short Answer (2 points each). Write your answer in the space provided below the question.

1. **According to Piaget, how do we know when the sensorimotor stage is coming to an end? That is, what developmental milestone marks the end of the sensorimotor stage and the beginning of the preoperational stage?**

Answer:

The development of symbolic thought, usually seen in early language.

2. **Define *therapeutic delay.***

Answer:

Psychiatric drugs change neurotransmitter activity fairly rapidly, in a few hours, but psychiatric symptoms can take weeks to abate. The time between effective modulation of transmitter activity and decrease of psychiatric symptoms is the therapeutic delay.

Part IV.

Longer Answer (4 points each). Write your answer in the space provided below the question.

1. **List and describe the four things that Piaget said that a child must have for cognitive development to proceed.**

Answer:

a. Biological maturation: Without sufficient development of the nervous system, cognitive development is impossible.
b. Physical experience: Children learn from interacting with the world, and nature itself provides rewards and punishments.
c. Social experience: Children can learn from interacting with other people, the most important of which are peers.
d. Adaptation: This comes in two forms, assimilation of new objects into existing ideas, and accommodation of ideas to deal with differences between new objects and old objects.

Piaget saw cognitive development as the result of an interaction between several factors, including other people, the physical world, and both the physiology and behavior of the child. Students need to list and describe these diverse elements.

A successful answer must include the four factors, with a brief description of each. Examples of each add weight to the description. The student should think of these factors as building on each other. Without biological maturation, nothing else is possible. With biological maturation, the child can now interact with the world and with other people. These interactions often take the form of adaptation. In this way, one factor builds on those before it in the list.

Students will have an easier time answering this question if they realize that the four factors described by Piaget are hierarchical to a certain extent. That is, each factor is required before the next factor is influential. This is most noticeable in the first three, since, for example, social experience may have much less effect in a child with significant neurological immaturity. This is true for both normal development and in cases of arrested neurological development. Students must see that this biological maturation is important in both normal and abnormal development.

2. A client's symptoms may reduce because we have given them an effective treatment, such as a drug. Describe two other reasons, neither of which is an actual effective treatment, why symptoms might decrease from the pretest to the posttest.

Answer:

Placebo effects: If the patient really believes that the treatment is effective, then it is more likely to change symptoms, even if the actual treatment is a sham.

Spontaneous remission: This is the "chance" category, in which symptoms reduce for no apparent reason. The patient's life may have changed between pretest and posttest (e.g., employment, relationships) and, if this is not known to the clinician, the change may be incorrectly attributed to the treatment.

Demand characteristics: The patient may have been lying at either the pretest or the posttest, or both. This lying may have been intentional or unintentional.

Regression toward the mean: The more extreme the symptoms on pretest, the less likely they are to be that extreme on the posttest. This is a statistical matter, and may be seen as a sort of spontaneous remission.

(Note that the question calls for only two responses; several more have been given here by way of illustration.)

There are many possible reasons for a change in symptoms over time. Such a change may be due to an effective treatment, but there are several other possible reasons for this change. Students must list and describe two of these, and providing real-world examples will help.

Students must demonstrate knowledge of the fact that symptoms can decrease for a number of different reasons. All of these may make the patient feel better, but not all of these are worth paying for. Students should be able to distinguish between the effect of successful intervention, and success that comes from other sources.

The first thing to think about is how much the patient knows about or expects from the treatment. The greater the expectation, the more likely is a placebo effect. Knowledge and compliance often lead to demand characteristics, and these may be known to the patient (intentional fraud), or they may be unknown to the patient (fooling themselves). The student must remember that symptoms can be affected by conscious or unconscious processes.

Think of the patient, a person who usually wants to get better. The more we want something, the more likely we are to think that we have found it, whether it is there or not. Some students will go one step further, and explain why simply making patients think that they are getting better is not enough. If it were, all we would need are more effective placebos. The top students see that treating the patient's impression of the symptoms is not as good an idea as treating the underlying problem.

2. **Draw and label a graph that shows habituation and dishabituation, in a study in which we try to see if a child can tell the difference between red and blue lights. Describe what is happening and what it means.**

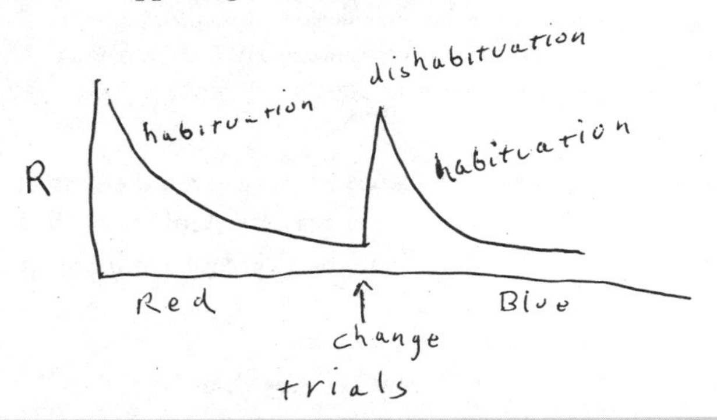

The student should draw a graph of responding as a function of trials. The abscissa will be divided into trials of red light presentation and trials of blue light presentation. The function will decrease quadratically with repeated stimulus presentation. When the stimulus is changed (as noted on the abscissa), the response will either recover, indicating dishabituation, or it will not, indicating a maintenance of habituation.

As in the midterm, students should be able to think in terms of a graph. This answer demonstrates the fundamentals of this type of nonassociative learning, and students might supplement this with another example from their own experience.

Students must demonstrate the knowledge that habituation involves a decrease in reactivity upon repeated presentation of a stimulus. They must also show a knowledge of the difference between habituation and dishabituation, both defined by response level. Some students will state that dishabituation is the test for habituation, as distinguished from adaptation or fatigue. Some will also state that dishabituation proves discrimination between the two stimuli, on some level, but that a failure to see dishabituation does not necessarily prove an absence of discrimination.

The best way to answer this question is to think of several examples of reduced reactivity upon repeated stimulus presentation. Many examples are discussed in class, and this helps students to think of examples from their own experience. Students should also remember that this nonassociative learning does not require conscious involvement. Students should remember that a change in stimulus often leads to a change in response, but only if the two stimuli can be discriminated. Students must also be able to think in terms of a graph, plotting reactivity as a function of stimulus presentation.

CHAPTER 28

WAYNE STATE UNIVERSITY

PSY 101: INTRODUCTORY PSYCHOLOGY

Mark A. Lumley, Associate Professor

FOUR EXAMS ARE GIVEN IN THE COURSE, EVENLY SPACED OVER THE SEMESTER, APPROXimately one every three and a half weeks. There is no formal midterm, and although exam #4 is given during final examination week, the exam is not cumulative. It covers material from the last quarter of the class only. I generally teach large introductory sections, approximately 200 students per class. I use multiple-choice tests exclusively, with 50 multiple-choice questions each.

For PSY 101 at Wayne State, there is an affiliated laboratory section, the grades of which go toward the final grade in the lecture. One-third of the final grade is from the lab, and the exams make up the remaining two-thirds. In recent semesters, I have counted the highest two of the first three exams given during the semester, and then required that the fourth exam be taken and counted toward the final grade. (Requiring the fourth makes sure that students who are satisfied with their first three scores do not skip the last quarter of the semester.) Thus, each exam counts for one-third of the two-thirds of the grade that comes from exams.

The objectives of the course are

- To expose the student to the wide range of topics studied by psychologists
- To teach critical thinking; that is, a skeptical yet humble approach to psychological information
- To teach ways to convert common sense into a research idea, and to test that idea empirically

- To have students learn some modicum of facts and theories from various domains of psychology
- To help students learn some living skills that may improve the quality of their own lives
- To challenge students on their extant beliefs regarding the degree to which various groups (e.g., gender, race, age) are mostly similar or mostly different
- To challenge students' often simplistic notion of the role of biology versus learning in particular behaviors (e.g., obesity, intelligence, mate preference, personality, disease)
- To help students become actively engaged with material, to help them become excited and curious, and to challenge their current beliefs

Attaining these objectives requires that students acquire as many of the following skills, habits of thought, and items of factual knowledge as possible:

- The ability to pose three questions whenever they encounter a psychological "fact":

1. What is the scientific evidence for this belief?
2. How good is that evidence?
3. What are alternative explanations?

- Some ability to differentiate the types of research designs that give rise to knowledge—for example, case study, survey, correlational, experimental, and so on
- The ability to differentiate correlation from causation
- Basic knowledge of the structure and function of the brain
- Basic knowledge of different learning theories and their applications to daily life

Of course, students also need to develop familiarity with all content in the text and as presented in lecture.

EXAM NO. 2

1. **The scientific study of sleep and dreams**
   a. started when CT and PET scans of the brain became possible.
   b. has been going on for centuries and strongly influenced Freud's theory of dreams.
   c. makes use of the EEG to measure muscle tension.
   d. was stimulated by the accidental discovery of rapid eye movement sleep in a child.

Answer: d

Answers a and b are simply factually wrong [note the adjective *scientific* in the question]. Answer c is tricky because it requires knowing that the EEG does not measure muscle tension (the EMG does). Answer d is correct, but this answer *seems* wrong because it includes reference to a child and was mentioned only in class. However, the instructor gave an engaging story about this historical event.

2. **After flying from California to New York, Arthur experienced a restless, sleepless night. His problem was most likely caused by a disruption of his normal**
   a. REM sleep.
   b. circadian rhythm.
   c. hypnogogic state.
   d. alpha wave pattern.

Answer: b

The direction of flight and name of person are irrelevant. Students need to be able to see jet lag and know what circadian rhythms are.

3. **When young adults live in a cave and are deprived of time cues, their sleeping and waking**
   a. follow a 23-hour cycle.
   b. follow a 24-hour cycle.
   c. follow a 25-hour cycle.
   d. lose all relationship to any time schedule.

Answer: c

This is a factual statement that could be retrieved through any of several associations.

4. **Jennifer has decided to go to bed early. Although her eyes are closed and she's very relaxed, she has not yet fallen asleep. Her EEG is most likely to indicate the presence of**
   a. delta waves.
   b. alpha waves.
   c. no muscle tension.
   d. rapid eye movements.

Answer: b

Factual question from text and noted in a classroom demonstration.

5. **Which of the following is *not* characteristic of REM sleep?**
   a. Voluntary muscles tense and become more active.
   b. Brain waves become more rapid.
   c. The eyes move rapidly under closed lids.
   d. The genitals are aroused.

Answer: a

All are true except a, which is backward. Muscles become more relaxed.

6. **Felicia goes to bed and will sleep for eight hours. During these hours, we can expect that _______ sleep will decrease and that _______ sleep will increase in duration.**
   a. stage 4; REM
   b. stage 2; stage 4
   c. stage 4; stage 3
   d. REM; stage 4

Answer: a

7. **Insomnia is a disorder involving**
   a. excessive use of sleeping pills or other drugs that induce sleep.
   b. recurring difficulty in falling or staying asleep.
   c. stopping breathing during sleep.
   d. the inability to sleep following a day of emotional stress or excitement.

Answer: b

Good example of two correct answers. Answer b is better than d, because it is more global. Answer d is correct also, but not in the presence of b, which is a better general definition.

8. **Mr. Oates sleeps restlessly, snorting and gasping all night. It is most likely that Mr. Oates suffers from**
   a. sleep apnea.
   b. narcolepsy.
   c. night terror.
   d. insomnia.

Answer: a

9. **Nightmares are found in _______, whereas night terrors occur during _______.**
   a. REM sleep; stage 4 sleep
   b. stage 1 sleep; REM sleep
   c. stage 4 sleep; stage 1 sleep
   d. stage 4 sleep; REM sleep

Answer: a

10. Which of the following is a scientific problem with Freud's theory of dreams?

a. People do not remember most of their dreams.
b. Males have erections during dreaming.
c. Humans dream in 90-minute cycles.
d. Many dreams seem boring.

Answer: c

Tricky question. Student must translate this into "which of these would Freud disagree with?"

11. A research study was presented on videotape in which a young woman slept for several days in the laboratory during the relaxed summer, and then again, right after studying very hard for examinations. The frequency of her eye movements during REM sleep was recorded during both occasions. What was found?

a. There was no difference in the frequency of her eye movements on the two occasions.
b. She had more eye movements during the time that she was studying.
c. She had more eye movements during the summer.
d. She had no eye movements at all.

Answer: b

Seems to most students a pure memory test based on a classroom video, yet, actually it is answerable from other information. Students already knew that REM sleep means eye movements (answer d is wrong). They should also surmise that no scientist presents results of a study where nothing differed (answer a is wrong). Finally, they should conclude from their knowledge of REM that it is linked with mental activity (answer b is right).

12. The hypothesis that dreams are random visual images generated by the brainstem and put together into a story is called the _______ hypothesis.

a. arousal-assembly
b. activation-symbiosis
c. generation-combination
d. activation-synthesis

Answer: d

When two answers have a similarity (b and d) and the others don't, chances are that one of the two is right.

13. While sleeping for eight hours at night, you are most likely to have a dream

a. during slow wave or deep sleep.
b. when you first start to fall asleep.
c. after six to eight hours of sleep.
d. within the first few hours of sleep.

Answer: c

14. Jet lag, problems with rotating shift work schedules, and "Sunday night insomnia" are all related to

a. minimal brain damage.
b. circadian rhythms.
c. inappropriate use of sleeping pills.
d. the 90-minute REM cycle.

Answer: b

This is almost the same question as question 2. Students should see if other questions will help them out.

15. Research on susceptibility to hypnosis indicates that

a. very few people can be hypnotized.
b. people who are most easily hypnotized usually have trouble paying attention to their own personal thoughts and feelings.
c. people who are highly hypnotizable tend to have rich fantasy lives.
d. nearly everyone can be hypnotized.

Answer: c

Purely factual and difficult only if you did not study well.

16. Just prior to awakening Michael from hypnosis, the therapist told him that during the next few days he would feel sick whenever he reached for a cigarette. Michael's therapist was making use of

a. post-hypnotic suggestion.
b. age regression.
c. a hidden observer.
d. hallucinogens.

Answer: a

17. In Aldous Huxley's *Brave New World,* infants develop a fear of books after books are repeatedly presented just before a loud noise. In this example, the loud noise is a(n)

a. unconditioned stimulus.
b. unconditioned response.
c. conditioned stimulus.
d. automatic reflex.

Answer: a

Here is a conceptual question, and students must not get caught up in all of the trappings around the basic model. This model was covered many times in class and applied to many things. This is a new application, and students need to see similarities with other applications (for example, Watson's conditioning of little Albert).

18. The infant Albert developed a fear of rats after a white rat was associated with a loud noise. In this example, fear of the white rat was the

a. UCS.
b. UCR.
c. CS.
d. CR.

Answer: d

Here, one is required to know not only the model parts, but also the abbreviations. Of course, most of them are in the preceding question. Students should use other questions to help them answer.

19. When you present a conditioned stimulus without an unconditioned stimulus, _______ will occur.

a. generalization
b. negative reinforcement
c. extinction
d. discrimination

Answer: c

20. Children who are taught to fear speeding cars may also begin to fear speeding trucks and motorcycles. This best illustrates:

a. latent learning.
b. secondary reinforcement.
c. shaping.
d. generalization.

Answer: d

21. The fact that people are more likely to fear insects, high places, and snakes rather than cars and electrical outlets is proposed as evidence for

a. biological preparedness in classical conditioning.
b. latent learning.
c. the equal ability of all stimuli to be conditioned.
d. operant reinforcement.

Answer: a

Items 19, 20, and 21 require purely factual knowledge.

22. You would be most likely to use operant conditioning to teach a dog to
a. fear cars in the street.
b. dislike the taste of dead birds.
c. salivate to the sound of a bell.
d. retrieve sticks and balls.

Answer: d

This question could be answered by simply trying to find out which of the four answers does not fit. It turns out that three are classical conditioning (reflex-based), and only one involves voluntary motor behavior.

23. What technique was used by the class to encourage one student to kiss another?
a. shaping
b. partial reinforcement schedules
c. secondary reinforcement
d. modeling

Answer: a

Here is a question designed to reward students who came to class. Those students who skipped would have no idea, as all four answers are legitimate.

24. Because Carol always picked up her newborn daughter when the baby cried, her daughter now cries all the time. In this case, picking up the infant served as a ________ for crying.
a. negative reinforcer
b. punisher
c. positive reinforcer
d. stimulus

Answer: c

Tricky, because picking up the infant could be either positive or negatively reinforcing, depending on the point of view. It is positive for the baby, but negatively reinforcing for the mother. Even if they get this part, students have trouble remembering what is the difference between positive and negative reinforcement (erroneously thinking the latter is punishment).

25. In operant conditioning, extinction occurs when
a. all animals of a species die.
b. you stop punishment.
c. you present the conditioned stimulus without the unconditioned stimulus.
d. you stop reinforcement.

Answer: d

Again, two answers are similar (b and d), just opposite. Choose one of the two.

26. Jennifer proofreads manuscripts for a publisher and is paid $10 for every three pages she reads. Jennifer is reinforced on a _______ schedule.

a. fixed-interval
b. fixed-ratio
c. variable-interval
d. variable-ratio

Answer: b

27. Negative reinforcers _______ the rate of responding and punishments _______ the rate of operant responding.

a. decrease; increase
b. increase; decrease
c. decrease; decrease
d. have no effect on; decrease

Answer: b

This question may help the student to answer question 24.

28. The use of physical punishment may

a. lead to the suppression but not the forgetting of undesirable behavior.
b. demonstrate that aggression is a way of coping with problems.
c. lead people to fear and avoid the punishing person.
d. do all the above.

Answer: d

I am willing to bet that if "all of the above" is included in exams, it is the right answer much more than 25 percent of the time!

29. Promising people rewards for doing what they already enjoy doing is likely to produce

a. overlearning.
b. the overjustification effect.
c. latent learning.
d. generalization.

Answer: b

Questions 29 through 36 are simply factual and largely definitional.

30. In a well-known experiment, nursery school children pounded and kicked a large inflated Bobo doll that an adult had just beaten on. This experiment served to illustrate the importance of

a. negative reinforcement.
b. operant conditioning.
c. respondent behavior.
d. observational learning.

Answer: d

31. After finding his friend's phone number, Alex was able to remember it only long enough to dial it correctly. In this case, the telephone number was clearly stored in his _______ memory.

a. iconic
b. short-term
c. flashbulb
d. long-term

Answer: b

32. Kathy performs better on her psychology exams if she studies the material 15 minutes every day for eight days than if she crams for two hours the night before the test. This illustrates what is known as

a. chunking.
b. the serial position effect.
c. mood-congruent memory.
d. the spacing effect.

Answer: d

33. Developed by the ancient Greeks, the method of loci is an example of

a. the spacing effect.
b. automatic processing.
c. a memory device.
d. the serial position effect.

Answer: c

34. Encoding information from the textbook into long-term memory is most successfully done by

a. effortful processing.
b. automatic processing.
c. serial positioning.
d. repression.

Answer: a

35. By flashing on a screen three rows of letters for only a fraction of a second, Sperling demonstrated that people have _______ memory.

a. echoic
b. iconic
c. state-dependent
d. implicit

Answer: b

36. Our immediate short-term memory for new material is limited to roughly _______ units of information.

a. 3
b. 7
c. 12
d. infinite

Answer: b

37. Creating and storing emotional memories involves all of the following brain parts *except* the

a. amygdala.
b. hippocampus.
c. cortex.
d. medulla.

Answer: d

The use of "except" language throws off some students. They need to think slowly and carefully. They also could think through the parts of the brain and recall that the medulla is far from the other structures.

38. Essay test questions measure _______; true-or-false questions measure _______.

a. recognition; relearning
b. free recall; relearning
c. free recall; recognition
d. relearning; free recall

Answer: c

39. When Loftus and Palmer asked observers of a filmed car accident how fast the vehicles were going when they "smashed" into each other, the observers developed memories of the accident that

a. omitted some of the most painful aspects of the event.
b. were more accurate than the memories of subjects who had not been immediately questioned about what they saw.
c. were influenced by whether or not Loftus and Palmer identified themselves as police officers.
d. portrayed the event as more serious than it had actually been.

Answer: d

40. A videotape shown in class showed a rat being injected with adrenaline prior to learning a maze. Adrenaline was found to

a. cause the fight-or-flight response.
b. improve memory.
c. interfere with memory.
d. have no effect.

Answer: b

Again, designed solely around a video shown in class, but actually mostly answerable if the student thinks through several things: "learning a maze" implies memory, and "have no effect" is always wrong (or the study would not be covered), so it comes down to b or c.

41. The idea that memories of childhood sexual abuse can be repressed

a. is not believed by most practicing therapists.
b. has been scientifically proven to exist.
c. received support from research that interviewed adults who were known to have been abused when they were children.
d. means that real sexual abuse is very rare.

Answer: c

A tough question that requires slowly thinking through all options.

42. The "eugenics" movement founded by Sir Francis Galton encouraged the

a. selective breeding of highly intelligent people.
b. creation of special education programs for intellectually inferior children.
c. construction of culturally and racially unbiased tests of intelligence.
d. use of factor analysis for identification of various types of intelligence.

Answer: a

43. Binet designed a test of intellectual abilities in order to
a. divide the races.
b. determine how much of intelligence is inherited and how much is learned.
c. identify children likely to have difficulty learning in school.
d. get his name on a famous intelligence test.
Answer: c

Here two of these answers are throwaways (a and d).

44. Tests designed to predict ability to learn new skills are called _______ tests.
a. achievement
b. interest
c. reliability
d. aptitude
Answer: d

45. If a test is standardized, this means that
a. it accurately measures what it is intended to measure.
b. a person's test performance can be compared with that of a large group of people.
c. most test scores will cluster near the average.
d. the test will yield consistent results when administered on different occasions.
Answer: b

46. If a test yields consistent results every time it is used, with very little error, it has a high degree of
a. standardization.
b. predictive validity.
c. reliability.
d. heritability.
Answer: c

47. During intelligence testing, 10-year-old Linda was found to have a mental age of a 12-year-old. According to the original IQ formula used by Terman, she would have an IQ score
a. of 120.
b. of 80.
c. in the mentally retarded range.
d. that was the same as the average 10-year-old.
Answer: a

48. Mr. and Mrs. Lembo are parents of a mentally retarded child. It is most likely that their child

a. is a female rather than a male.
b. suffers obvious physical defects.
c. was born with an extra chromosome.
d. will have difficulty adapting to the normal demands of independent adult life.

Answer: d

This question has three specific answers and one general one (d). Usually general answers are more correct than specific ones.

49. Which of the following observations provides the best evidence that intelligence test scores are influenced by heredity?

a. Japanese children have higher average intelligence scores than American children.
b. Fraternal twins are more similar in their intelligence scores than are ordinary siblings.
c. Identical twins reared separately are more similar in their intelligence scores than fraternal twins reared together.
d. The intelligence scores of children are positively correlated with the intelligence scores of their parents.

Answer: c

50. What is the most common cause of mental retardation?

a. trisomy 21
b. environmental deprivation
c. Down syndrome
d. fetal alcohol syndrome

Answer: d

EXAM NO. 4

Multiple-choice questions are notoriously difficult to construct to test how well students understand concepts. Also, the students whom I teach vary widely in their intellect and readiness for college. The exams that I construct, therefore, probably rely more heavily on factual information (basic knowledge) than on conceptual applications. I find that students routinely perform poorly on conceptual questions, and to help them feel some degree of success on the exam, I use primarily factual questions.

Nevertheless, learning is an active process, not just a matter of memorization of facts. You must work at studying a textbook and actively think while in lecture. Too many students read a textbook—especially the well-written, user-friendly introductory psychology texts—as if they were reading an enjoyable magazine; that is, they do not *actively* study. Similarly, many students attend

lecture passively, either "enjoying a good talk," transcribing everything the instructor says, or simply trying to stay awake.

When studying the text, students need to follow some studying protocol, such as SQ3R: Survey, Question, Read, Respond, Review. In general, they need to read much more slowly, pausing often to think, question, and link concepts. Also, a great deal of marker on a page is a clear sign that the student is not critically thinking as he or she reads; it is evidence that he or she is having trouble differentiating the more important from the less important.

When attending lecture, students would do well to have at least previewed the text, to see if they can"think ahead" of the instructor. I would prefer that every student audiotape the lecture, and then listen actively (without transcribing) during the lecture—again, anticipating, linking, questioning, and so on. They can take notes later from the audiotape.

Most instructors will give away or hint about information that will be tested. You pick up on these cues if you just actively listen. For example, the instructor's saying something twice, slowing down to emphasize a point, and spending a lot of time on a topic are common signs that the instructor will revisit this material on a test.

One of my favorite phrases, which, unfortunately, captures the essence of many lectures is: "A lecture is too often defined as the process by which information is transferred from the notes of one person to the notes of another person without going through the minds of either."

Get a jump on the material. Spaced practice is more successful than massed practice. A little studying each day is more successful than a full night of studying before an exam.

Select the one best answer for each question and mark that letter on your Scantron.

1. A stressor is a(n)
   a. physiological response to negative life events.
   b. competitive, hard-driving, impatient person.
   c. environmental event that threatens or challenges us.
   d. exercise program designed to increase our ability to handle normal stress.

Answer: c

2. **The inner part of the adrenal gland secretes the hormone _______, and the outer part of the adrenal gland secretes _______.**
   a. norepinephrine (noradrenaline); growth hormone
   b. testosterone; epinephrine (adrenaline)
   c. cortisol; insulin
   d. epinephrine (adrenaline); cortisol

Answer: d

The preceding two questions require the students to know both answers, because either a or d is accurate for the first blank.

3. The general adaptation syndrome describes stages in the

a. conditioning of the immune response.
b. body's response to exercise.
c. body's response to prolonged stress.
d. process of biofeedback.

Answer: c

4. Research has indicated that rats and monkeys are more likely to develop ulcers when exposed to _______ shock.

a. uncontrollable
b. predictable
c. low intensity
d. electro-

Answer: a

5. Friedman and Rosenman referred to noncompetitive, relaxed, and easygoing individuals as _______ personalities.

a. extroverted
b. health-prone
c. type A
d. type B

Answer: d

Students can benefit from examining other answers, such as number 6, to help out with this one.

6. What component or aspect of the type A behavior pattern is thought to contribute to heart disease?

a. hostility
b. competitiveness
c. time urgency (doing things in a hurry)
d. trying to climb the corporate ladder

Answer: a

7. Which of the following best explains why stress increases a person's vulnerability to infections by bacteria and viruses?

a. Stress hormones reduce the activity of lymphocytes.
b. Stress hormones increase the depositing of cholesterol and fat around the heart.
c. Stress hormones speed up "hardening" of the arteries.
d. Stress hormones trigger release of digestive acids.

Answer: a

This one requires that the student understand both stress and the body's defenses against pathogens (i.e., the immune system).

8. In researching taste aversion in rats, Ader and Cohen discovered that sweet-tasting water was a conditioned stimulus for
   a. the release of pain-killing endorphins.
   b. the suppression of the immune system.
   c. an overproduction of acetylcholine.
   d. overeating.

Answer: b

9. Pennebaker's research on writing about traumatic events shows that
   a. people who write about trivial or superficial topics have health benefits.
   b. people really need to write for many months or years to overcome traumas.
   c. writing privately for a few days can lead to health benefits.
   d. writing is as helpful as having psychotherapy.

Answer: c

10. Electronically recording, amplifying, and displaying information about a person's physiological responses is called
   a. stress management.
   b. biofeedback.
   c. relaxation training.
   d. behavioral medicine.

Answer: b

11. Research on nicotine and smoking has shown that
   a. many people start smoking late in life.
   b. nicotine is not addictive; only the act of smoking is.
   c. stopping smoking leads to a withdrawal syndrome.
   d. there is no genetic predisposition to becoming addicted to cigarettes.

Answer: c

12. Psychologists are most likely to define someone's behavior as a disorder if it is
   a. unloving and prejudicial.
   b. biologically based and habitual.
   c. unconsciously motivated.
   d. unusual and socially unacceptable.

Answer: d

13. DSM-IV is a system that is widely used for
   a. identifying the causes of psychological abnormality.
   b. distinguishing sanity from insanity.
   c. treating depression.
   d. classifying psychological disorders.

Answer: d

14. Psychological disorders in which people lose contact with reality and experience irrational ideas and distorted perceptions are known as _______ disorders.
   a. psychotic
   b. generalized anxiety
   c. dissociative
   d. obsessive-compulsive

Answer: a

15. An episode of intense fear or dread that lasts for several minutes and is accompanied by shortness of breath, trembling, dizziness, choking, or heart palpitations is called a(n)
   a. phobic disorder.
   b. panic attack.
   c. obsessive-compulsive disorder.
   d. heart attack.

Answer: b

16. A person who has agoraphobia is most likely to
   a. avoid dust and dirt.
   b. stay away from fire.
   c. avoid empty rooms.
   d. stay close to home.

Answer: d

> **Here the student must be able to recall the characteristics of a disorder, in contrast to the prior questions, in which the characteristics were given and the student needed to know the label for the disorder.**

17. Mrs. Swift is alarmed because she has repeated and irrational thoughts of murdering her young children. Her experience is a good example of a(n)

a. delusion.
b. phobia.
c. obsession.
d. compulsion.

Answer: c

Here knowing the difference between two frequently confused phenomena, obsessions and compulsions, is needed.

18. During a stressful military battle, Fong suddenly went blind. When he was hypnotized by an army psychologist, however, Fong's eyesight returned completely. Fong apparently suffered from

a. hypochondriasis.
b. a dissociative disorder.
c. a conversion disorder.
d. a generalized anxiety disorder.

Answer: c

19. Most people with multiple personality disorder

a. have both auditory and visual hallucinations.
b. had a loving, stress-free childhood.
c. had brain damage as infants.
d. are highly hypnotizable.

Answer: d

20. What disorder is so common that it is called the "common cold" of psychological disorders?

a. depression
b. schizophrenia
c. antisocial personality
d. low self-esteem

Answer: a

21. In which disorder do people alternate between states of lethargic or weary hopelessness and wild overexcitement?

a. multiple personality disorder
b. bipolar disorder
c. obsessive-compulsive disorder
d. schizophrenia disorder

Answer: b

22. When people experience a failure or bad event, they are most likely to become depressed if they attribute or explain the cause of the failure as due to

a. racism, sexism, or ageism.
b. God.
c. something that is internal, stable, and global about themselves.
d. a neurochemical imbalance.

Answer: c

23. Seligman's research found that if dogs were shocked in a box from which they could not escape, they developed a condition that looked like the disorder of _______ in people. He called this phenomenon _______.

a. depression; learned helplessness
b. depression; neurotic conflict
c. schizophrenia; learned helplessness
d. schizophrenia; shock trauma

Answer: a

24. Seeing one-eyed monsters would be a(n) _______. Believing that you are Abraham Lincoln would be a _______.

a. delusion; compulsion
b. obsession; delusion
c. hallucination; compulsion
d. hallucination; delusion

Answer: d

Much of abnormal psychology is purely factual and often definitional.

25. Schizophrenia is associated with an excess or overactivity of which neurotransmitter?

a. norepinephrine
b. dopamine
c. serotonin
d. acetylcholine

Answer: b

26. Research on the causes of schizophrenia indicate that

a. there is a genetic predisposition to schizophrenia.
b. almost anybody will develop schizophrenia if exposed to extensive environmental stress.
c. children adopted by a schizophrenic parent are likely to develop schizophrenia themselves.
d. disordered communication in the family causes schizophrenia.

Answer: a

27. There is research evidence that people with _______ have a reduced level of fear and sympathetic arousal when they are threatened.

a. hypochondriasis
b. schizophrenia
c. antisocial personality disorder
d. multiple personality disorder

Answer: c

28. Psychoanalytic techniques are designed primarily to help clients

a. focus on their immediate conscious feelings.
b. feel more trusting toward others.
c. become more insightful about repressed conflicts and impulses.
d. develop greater self-esteem.

Answer: c

Although all of these answers are somewhat true, answer c is the most applicable to psychoanalytic therapy, whereas the others apply to other types of therapies.

29. When the therapist encouraged Jeff to talk about his resentment toward his wife, Jeff quickly changed the subject by telling a joke. To a psychoanalyst, Jeff's behavior illustrates

a. resistance.
b. transference.
c. fixation.
d. free association.

Answer: a

30. Mr. Phillips has recently begun to express feelings of hostility and resentment toward his therapist, who is actually a caring and helpful woman. A psychoanalyst would most likely consider Mr. Phillips' hostility toward his therapist to be an example of

a. transference.
b. repression.
c. the placebo effect.
d. sublimation.

Answer: a

31. Which approach emphasizes the importance of providing clients with feelings of unconditional acceptance?

a. rational-emotive therapy
b. psychoanalysis
c. client-centered or person-centered therapy
d. systematic desensitization

Answer: c

32. Psychological research on the principles of learning or conditioning has most directly influenced the development of

a. behavior therapy.
b. person-centered therapy.
c. psychoanalysis.
d. rational-emotive therapy.

Answer: a

33. To help Michael overcome his fear of taking tests, his therapist instructs him to relax and then to imagine taking a quiz. The therapist is using

a. psychoanalysis.
b. person-centered therapy.
c. systematic desensitization.
d. aversive conditioning.

Answer: c

34. A treatment program in which an alcoholic client drinks alcohol that contains a drug that makes the person sick is using a technique known as

a. operant conditioning.
b. free association.
c. systematic desensitization.
d. aversive conditioning.

Answer: d

35. A token economy, in which patients in a hospital are given points for good behavior and lose points for bad behavior is based on the principles of

a. operant conditioning.
b. systematic desensitization.
c. classical conditioning.
d. observational learning.

Answer: a

36. It is totally stupid for you to think you are worthless because your parents criticize! You're worthless only if you think you are. This statement would most likely be made to a client by a _______ therapist.

a. rational-emotive
b. psychoanalytic
c. person-centered
d. behavior

Answer: a

37. Cognitive therapists such as Aaron Beck are most likely to encourage depressed clients to

a. test whether their negative beliefs about themselves are true.
b. confront their parents with their anger.
c. undergo hypnosis to learn more about the cause of their depression.
d. get more active and do more rewarding activities during the day.

Answer: a

Most of these questions require that the student be familiar with various forms of psychotherapy and examples of what they look like in practice.

38. Gretchen avoids shaking people's hands or touching doorknobs, because she is afraid of contracting infectious diseases. When she does shake hands or touch knobs, she washes her hands for about 15 minutes. Research suggests that an especially effective treatment for her difficulty would involve

a. psychoanalysis.
b. rational-emotive therapy.
c. flooding and response prevention.
d. electroconvulsive therapy.

Answer: c

This question requires knowledge both of the diagnosis and of the specific treatment.

39. The most effective psychotherapists are those who

a. employ personality tests to diagnose their clients' difficulties.
b. use a wide variety of therapeutic techniques.
c. have had many years of experience practicing psychotherapy.
d. establish an empathic, caring relationship with their clients.

Answer: d

Whereas all of these answers seem true, the central theme about efficacy in therapy leads one to choose d as the best answer.

40. Although Dr. Miller prescribes medication for the treatment of chronic depression, she encourages relaxation training for clients suffering from excessive anxiety. It is most likely that Dr. Miller is a

a. psychoanalyst.
b. Gestalt therapist.
c. behavior therapist.
d. psychiatrist.

Answer: d

Here the student must be able to refrain from selecting the "behavior therapist," because of the seemingly unimportant statement about medications, which actually indicates that Dr. Miller must be a physician.

41. Which drugs appear to produce their effects by blocking receptor sites for dopamine?

a. antipsychotic drugs
b. antidepressant drugs
c. antianxiety drugs
d. pain relievers

Answer: a

42. Who is most likely to benefit from a drug that increases serotonin levels?

a. someone who feels helpless and apathetic and thinks her life is meaningless and worthless
b. someone who hears imaginary voices telling her that she will suffer a fatal accident
c. someone who is so addicted to cigarettes that she now worries about her health
d. someone who has lost his sense of identity and wandered from his home to a distant city

Answer: a

This one requires the student to know both what class of drugs have serotonergic agonist properties and how to pick up the diagnosis of depression.

43. The potential benefits of electroconvulsive therapy were suggested by the observation that

a. shocking someone usually got the person very angry.
b. drugs often were not effective, so shock therapy might be.
c. dogs stuck in a box that gives electric shocks develop depression.
d. people who have seizures typically do not have major depression.

Answer: d

44. We have a tendency to explain the behavior of others in terms of _______ and to explain our own behavior in terms of _______.

a. informational influence; normative influence
b. the situation; personality traits
c. normative influence; informational influence
d. personality traits; the situation

Answer: d

45. The "foot-in-the-door" phenomenon refers to the tendency to

a. neglect critical thinking because of a strong desire for social harmony within a group.
b. perform simple tasks more effectively in the presence of others.
c. agree to a large request if one has previously agreed to a small request.
d. lose self-control in group situations where one might be anonymous.

Answer: c

46. Fernando used to be in favor of capital punishment before he voluntarily offered arguments opposing it in a college debate class. He now has a negative attitude toward capital punishment. His change in attitude is best explained by _______ theory.

a. cognitive dissonance
b. reactance
c. scapegoat
d. self-perception

Answer: a

47. Solomon Asch reported that people conformed to a group's judgment of the lengths of lines

a. even when the group judgment was clearly incorrect.
b. only when the group was composed of at least six members.
c. even when group members were uncertain and not unanimous in their decisions.
d. only when members of the group were of high status.

Answer: a

This particular question tripped up most students. The answers were too confusing and will be changed next time.

48. Most people are likely to be surprised by the results of Milgram's obedience experiment because

a. the "learners" made so few learning errors under stressful circumstances.
b. the "teachers" actually enjoyed shocking another person.
c. the "learners" obediently accepted painful shocks without any protest.
d. the "teachers" were more obedient than most people would have predicted.

Answer: d

49. According to Milgram, the most fundamental lesson to be learned from his study of obedience is that

a. people are naturally hostile and aggressive.
b. even ordinary people, who are not usually hostile, can become destructive or violent.
c. only disordered or unusual people will do violent or evil actions.
d. people value freedom and will not be forced into doing things they do not believe in.

Answer: b

This question requires that the student get the main point of a study. This is a highly conceptual question that often throws students who only recall what a study was about.

50. When a PSY 101 student had a flat tire on a lonely, mostly empty road in the country, a passerby helped in less than 10 minutes. One year later, the same student had a flat tire on a busy freeway, and it was an hour before someone finally stopped to offer help. This student's experience best illustrates

a. the fundamental attribution error.
b. the mere exposure effect.
c. the "I don't know how to avoid sharp objects" effect.
d. the bystander effect.

Answer: d

PART FOUR

FOR YOUR REFERENCE

CHAPTER 29

A GLOSSARY OF PSYCHOLOGY

Association Mental link between stimuli, thoughts, or memories.

Attitude An enduring idea concerning people or things; in contrast to a *belief,* it has an emotional component.

Autonomic nervous system (ANS) The part of the *peripheral nervous system (PNS)* that transmits nerve impulses to and from internal organs.

Behavior Anything an organism does. See *covert behavior* and *overt behavior.*

Belief An enduring idea about people or things; in contrast to *attitude,* it has no emotional component.

Catharsis Psychoanalytic term for a release of emotion (through words or action) that relieves underlying psychic pressures.

Central nervous system (CNS) Nervous tissue of the spinal cord and brain.

Cerebral cortex Convoluted outer layer of the brain that is the site of higher mental processing.

Chromosomes The cellular structures that carry genes.

Classical conditioning Learning based on associations between stimuli; first extensively studied by Ivan Pavlov.

Cognition Conscious thought and thinking.

Confounding An experimental or observational situation in which two or more causes of a result cannot be separated.

Conscious mind Thinking and awareness in the present.

Contingency Term in *operant conditioning* for the relationship between a behavior and its consequences.

Correlation Statistical operation by which two or more *variables* are compared to see how they correspond.

Covert behavior Internal mental processes not objectively observable. Many aspects of thought and emotion are covert. Contrast *overt behavior.*

Critical period In development, a time range during which a given behavior must be acquired or it will never be acquired. Also see *sensitive period.*

Cross-sectional research Research approach that simultaneously studies two groups of subjects, usually of different ages or in different developmental periods.

Delusion A belief without basis in reality. Compare *hallucination.*

Dissociation In *hypnosis,* a state in which awareness is separated from reality or in which thoughts and feelings are separated from each other. In mental disorders, a condition in which aspects of personality, emotion, and memory are separated from each other.

DNA (deoxyribonucleic acid) A nucleic acid containing deoxyribose sugar, DNA carries genetic information and is capable of self-replication as well as synthesis of RNA.

Drive A psychological state arising from a need.

EEG (electroencephalograph) Device that monitors and records global brain activity.

Emotion Physiological arousal accompanied by cognitive interpretation.

Empirical Observable.

Endorphin A neurotransmitter chemical that functions as a natural painkiller.

Environment As used in psychology, any external conditions that affect an organism from conception on.

Factor analysis Correlational statistical operation to compare variables in order to determine how they interrelate and overlap.

Gene Hereditary unit on a specific chromosome location; determines a specific trait of an organism.

Genotype All the genetic traits of an organism, whether or not these are expressed as phenotypes.

Habit Highly repetitive behavior.

Hallucination *Percept* without basis in reality.

Homeostasis Tendency of organisms to establish and maintain a balance between internal systems and external environment.

Hormone Substance secreted directly into the bloodstream, which affects a target organ or organs.

Hypnosis A sleeplike state, usually induced in one person by another, in which the subject may be suggestible.

Hypothalamus Brain structure involved with homeostatic functions.

Hypothesis A falsifiable statement about potential outcome or causation.

Information processing A model of thinking and memory expressed in terms of sequential operations; borrows extensively from computer science.

Instinct A built-in behavior that occurs in specific situations.

Introspection Mental processes of looking within oneself and describing thoughts and emotions.

IQ (intelligence quotient) As originally conceived, the ratio of test performance (mental age) to chronological age. Now defined as the ratio of a subject's test performance to the performance of others in the same age range.

Longitudinal study A research study conducted over time.

Long-term memory (LTM) Relatively permanent memory accumulated and stored over the life of the individual.

Motive Cognitive state based on needs, drives, or other causes, in which an organism seeks satisfaction.

Need Physiological depletion giving rise to *drives* or *motives.*

Nerve A bundle of *neurons.*

Neuron Nerve cell.

Neurotic Outmoded (but common) term for mental disorders that impinge on a person's life without rendering the person wholly dysfunctional.

Neurotransmitters Chemical substance released at the ends of neurons by which nerve impulses are transmitted.

Norm Used developmentally, the age at which a given behavior usually occurs. Used in psychological testing, the statistically established basis for interpreting an individual's performance. Used in abnormal or social psychology, behavior considered acceptable and appropriate by society.

Observational learning Learning based on observing others and how they are rewarded or punished.

Operant conditioning Learning based on consequences.

Overt behavior Observable and measurable behavior; contrast *covert behavior.*

Parasympathetic nervous system (PNS) Part of the *autonomic nervous system (ANS)* that returns the organism to normal functioning after action by the *sympathetic nervous system.*

Percept A given instance of perception.

Perception Recognition and/or interpretation of stimuli.

Peripheral nervous system (PNS) Sensory and motor nervous tracts that transmit impulses to and from the *central nervous system (CNS).*

Phenomenology Philosophical and psychological approach assuming that each individual must be understood from his or her own point of view.

Phenotype The visible traits of an organism.

Population In biology, the members of a species in a given area that can breed with one another. In experimental psychology, all of a designated type of subject.

Psychoanalysis General term for Sigmund Freud's theory of mind and the clinical procedures for understanding these processes and treating mental disorders.

Psychodynamic Adjective used to describe theories growing out of *psychoanalysis,* but extending or differing from it.

Psychomotor Movement and the cognitive activity directing it.

Psychotic Adjective applied to mental disorders rendering a person thoroughly dysfunctional (and potentially dangerous to self or others).

Punishment Suppression of a behavior due to unfavorable consequences. Negative punishment occurs when an appetitive stimulus is removed as a consequence of a behavior. Positive punishment occurs when an aversive stimulus occurs as a consequence of a behavior. Compare *reinforcement.*

Quasi-experiment A procedure that resembles an experiment, except that researchers must work with naturally occurring conditions over which they have little or no control.

Rapid eye movement (REM) Eye movement during sleep, especially dreaming.

Reflex Built-in behavior minimally under voluntary control.

Reinforcement Strengthening a behavior through favorable consequences. Negative reinforcement occurs when an aversive stimulus is removed as a consequence of a behavior. Positive reinforcement occurs when an appetitive stimulus comes as a consequence of a behavior.

Reliability Extent to which a test yields similar results on different occasions.

Schema Cognitive structure providing a framework for perception and knowledge.

Sensitive period Developmental time range during which a behavior can be acquired most efficiently. Compare *critical period.*

Short-term memory (STM) Temporary memory, analogous to RAM in a computer. Compare *long-term memory (LTM).*

Standardization In psychological testing, the procedures that establish a basis for interpreting test scores' mean.

Stereotype In social psychology, attitude or belief concerning characteristics thought to apply to all members of a given group.

Stimulus Sensory event the organism is capable of detecting.

Subconscious mind Cognitive functioning we normally are not aware of, but which we can be readily made aware of. Compare *unconscious mind.*

Sympathetic nervous system Part of the *autonomic nervous system (ANS)* that reacts to stress and emergency situations, preparing the organism for flight or fight.

Synapse The structure at the juncture of two neurons.

Temperament Personality characteristics more or less present at birth.

Trait Enduring aspect of a physical or psychological makeup.

Unconscious mind Mental processes to which we normally have no access and of which we are unaware, but which nevertheless affect us.

Validity Extent to which a test measures what it is intended to measure.

Variable Any measurable factor that can differ along any dimension.

CHAPTER 30

RECOMMENDED READING

Baron, Robert A., and Michael J. Kalsher. *Psychology.* Allyn and Bacon, 1995.

Butler, Gillian, and Freda McManus. *Psychology: A Very Short Introduction.* Oxford University Press, 1998.

Davis, Stephen F., and Joseph J. Palladino. *Psychology.* Macmillan, 1995.

Erikson, Erik H. *Childhood and Society.* W. W. Norton, 1993.

_____. *Identity and Life Cycle.* W. W. Norton, 1994.

Fernald, L. Dodge. *Psychology.* Prentice Hall, 1996.

Fogiel, M. *The Psychology Problem Solver.* Research and Education Association, 1990.

Freud, Sigmund. *The Basic Writings of Sigmund Freud.* Modern Library, 1995.

_____. *Introductory Lectures on Psychoanalysis.* Liveright, 1989.

Gleitman, Henry. *Basic Psychology.* W. W. Norton, 1995.

Goldstein, E. Bruce. *Psychology.* Brooks/Cole, 1994.

Hilgard, Richard C. *Hilgard's Introduction to Psychology.* HBJ College and School Division, 1998.

Huffman, Karen, et al. *Essentials of Psychology in Action.* John Wiley, 1994.

Jung, Carl Gustav. *The Portable Jung.* Viking, 1976.

Kalat, James W. *Introduction to Psychology.* Brooks/Cole, 1996.

Kassin, Saul M. *Psychology.* Prentice Hall, 1997.

Lefton, Lester A. *Psychology.* Allyn and Bacon, 1996.

Maslow, Abraham H. *Motivation and Personality.* Addison-Wesley, 1987.

_____. *Toward a Psychology of Being.* John Wiley, 1968.

Mealey, Donna L., et al. *Studying for Psychology.* Addison-Wesley, 1996.

Morris, Charles G., and Albert A. Maisto. *Psychology: An Introduction.* Prentice Hall, 1998.

Myers, David G. *Exploring Psychology.* Worth, 1998.

Parrott, Leslie. *How to Write Psychology Papers.* Addison-Wesley, 1996.

Piaget, Jean. *The Essential Piaget.* Jason Aronson, 1995.

Reber, Arthur S. *The Penguin Dictionary of Psychology.* Penguin USA, 1996.

Rogers, Carl R. *Client-Centered Therapy: Its Current Practice, Implications, and Theory.* Constable, 1995.

_____. *A Way of Being.* Houghton Mifflin, 1995.

Scott, Jill. *The Psychology Student Writer's Manual.* Prentice Hall, 1998.

Seamon, John G., and Douglas T. Kendrick. *Psychology.* Prentice Hall, 1994.

Skinner, B. F. *About Behaviorism.* Random House, 1976.

_____. *Science and Human Behavior.* Free Press, 1965.

_____. *Walden Two.* Allyn and Bacon, 1976.

Stanovich, Keith E., ed. *How to Think Straight about Psychology.* Addison-Wesley, 1997.

Wade, Carole, and Carol Tarvis. *Invitation to Psychology.* Longman, 1998.

Weber, Ann L. *Introduction to Psychology* (HarperCollins College Outline Series). HarperCollins, 1991.

Wittig, Arno F. *Introduction to Psychology* (McGraw-Hill College Review Books Series). McGraw-Hill, 1989.

Woods, Paul J., ed. *Is Psychology the Major for You? Planning for Your Undergraduate Years.* American Psychological Association, 1987.

Zimbardo, Philip G., and Ann L. Weber. *Psychology.* Addison-Wesley, 1997.

Note: This index covers Part One: Preparing Yourself and Part Two: Study Guide, pages VII-149 of the text. Material in the sample exams, pages 152–329, is not indexed.